Cookies

Systems of Equations and Linear Programming

Teacher's Guide

This material is based upon work supported by the National Science Foundation under award numbers ESI-9255262, ESI-0137805, and ESI-0627821. Any opinions, findings, and conclusions or recommendations expressed in this publication are those of the authors and do not necessarily reflect the views of the National Science Foundation.

Key Curriculum
1150 65th Street
Emeryville, California 94608
email: editorial@keypress.com
www.keycurriculum.com

First Edition Authors

Dan Fendel, Diane Resek, Lynne Alper, and Sherry Fraser

Contributors to the Second Edition

Sherry Fraser, Jean Klanica, Brian Lawler, Eric Robinson, Lew Romagnano, Rick Marks, Dan Brutlag, Alan Olds, Mike Bryant, Jeri P. Philbrick, Lori Green, Matt Bremer, Margaret DeArmond

Project Editor

Sharon Taylor

Consulting Editor

Mali Apple

Project Administrator

Juliana Tringali

Professional Reviewer

Rick Marks, Sonoma State University, CA

Calculator Materials Editor

Josephine Noah

Math Checker

Carrie Gongaware

Production Editor

Andrew Jones

Production Director

Christine Osborne

Executive Editor

Josephine Noah

Mathematics Product Manager

Tim Pope

Publisher

Steven Rasmussen

Contents

Blackline Masters

Calculator Guide and Calculator Notes

Introduction

Cookies Unit Overview

Intent

The focus of this unit is the use of graphs of linear equations and inequalities to analyze and solve problems. The overarching goal of the unit is for students to deepen their understanding of the relationship between equations or inequalities and their graphs and to reason about and solve problems both symbolically and graphically.

The central unit problem involves a bakery that produces two kinds of cookies. The bakery faces constraints on ingredients, preparation time, and oven space. The two kinds of cookies require different amounts of ingredients and earn different amounts of profit. The task for students is to determine how many of each kind of cookie the bakery should produce to maximize overall profit.

Mathematics

The central mathematical focus of *Cookies* is the formulation and solution of problems of optimization, or linear programming problems. In problems of this type, a linear function is to be optimized and a set of linear conditions constrains the possible solutions. Linearity is an important feature of these two-variable problems, in two ways:

- The constraints are linear, so the feasible region is a polygon and its vertices can be found by solving pairs of linear equations.

- The expression to be maximized or minimized is linear, so the points that give this expression a particular value lie on a straight line, and investigating a series of values produces a family of parallel lines.

The linear programming problems that students encounter in this unit involve only two variables and a limited number of constraints. Their solutions are therefore easier to understand graphically, and the algebra needed to find their exact solutions is manageable.

The main concepts and skills that students will encounter and practice during the unit are summarized here.

Using Variables to Represent Problems

- Expressing and interpreting constraints using inequalities

- Expressing problem situations using systems of linear equations

Working with Variables, Equations, and Inequalities

- Finding equivalent equations and inequalities

- Solving linear equations for one variable in terms of another

- Developing and using a method for solving systems of two linear equations in two unknowns

- Recognizing inconsistent systems and dependent systems

Graphing

- Graphing linear inequalities and systems of linear inequalities

- Finding the equation of a straight line and the inequality for a half plane

- Using graphing calculators to draw feasible regions

- Relating the intersection point of graphed lines to the common solution of the related equations

- Using graphing calculators to estimate coordinates of points of intersection

Reasoning Based on Graphs

- Recognizing that setting a linear expression equal to a series of constants produces a family of parallel lines

- Finding the maximum or minimum of a linear equation over a region

- Examining how the parameters in a problem affect the solution

- Developing methods of solving linear programming problems with two variables

Creating Word Problems

- Creating problems that can be solved using two equations in two unknowns

- Creating problems that can be solved by linear programming methods

Progression

Following an introduction to the unit problem, students explore the graphing of linear inequalities and create graphs of feasible regions. Using graphical analysis, they maximize and minimize profit and cost functions and learn that the best answers for linear programming problems are at the vertices of the feasible regions. This leads to a focus on methods for solving linear systems. The unit concludes with students creating their own linear programming problems.

Cookies and Inequalities

Picturing Cookies

Using the Feasible Region

Points of Intersection

Cookies and the University

Creating Problems

Supplemental Activities

Unit Assessments

Pacing Guides

50-Minute Pacing Guide (29 days)

Day	Activity	In-Class Time Estimate
Cookies and Inequalities		
1	*How Many of Each Kind?*	45
	Homework: *A Simpler Cookie*	5
2	Discussion: *A Simpler Cookie*	15
	How Many of Each Kind? (continued)	30
	Homework: *Investigating Inequalities*	5
3	Discussion: *Investigating Inequalities*	20
	My Simplest Inequality	25
	Homework: *Simplifying Cookies*	5
4	Discussion: *Simplifying Cookies*	15
Picturing Cookies		
	Picturing Cookies—Part I	30
	Homework: *Inequality Stories*	5
5	Discussion: *Inequality Stories*	15
	Picturing Cookies—Part I (continued)	20
	Introduce: *POW 4: A Hat of a Different Color*	10
	Homework: *Healthy Animals*	5
6	Discussion: *Healthy Animals*	15
	Picturing Cookies—Part II	30
	Homework: *What's My Inequality?*	5

7	Discussion: *What's My Inequality?*	25
	Feasible Diets	20
	Homework: *Picturing Pictures*	5
8	Discussion: *Picturing Pictures*	20

Using the Feasible Region

	Profitable Pictures	30
	Homework: *Curtis and Hassan Make Choices*	0
9	Discussion: *Curtis and Hassan Make Choices*	15
	Profitable Pictures (continued)	30
	Homework: *Finding Linear Graphs*	5
10	Discussion: *Finding Linear Graphs*	15
	Presentations: *POW 4: A Hat of a Different Color*	15
	Introduce: *POW 5: Kick It!*	10
	Hassan's a Hit!	10
	Homework: *You Are What You Eat*	0
11	Discussion: *You Are What You Eat*	20
	Hassan's a Hit (continued)	30
	Homework: *Changing What You Eat*	0
12	Discussion: *Changing What You Eat*	40
	Homework: *Rock 'n' Rap*	10
13	Discussion: *Rock 'n' Rap*	20
	A Rock 'n' Rap Variation	30
	Homework: *Getting on Good Terms*	0
14	Discussion: *Getting on Good Terms*	15

	Rock 'n' Rap on the graphing calculator	30
	Homework: *Going Out for Lunch*	5
15	Discussion: *Going Out for Lunch*	20

Points of Intersection

	Get the Point	25
	Homework: *Only One Variable*	5
16	Discussion: *Only One Variable*	25
	Get the Point (continued)	20
	Homework: *Set It Up*	5
17	Discussion: *Set It Up*	25
	Get the Point (continued; presentations)	20
	Homework: *A Reflection on Money*	5
18	Discussion: *A Reflection on Money*	20
	Presentations: *POW 5: Kick It!*	20
	Introduce: *POW 6: Shuttling Around*	10
	Homework: *More Linear Systems*	0
19	Discussion: *More Linear Systems*	20

Cookies and the University

	"How Many of Each Kind?" Revisited	25
	Homework: *A Charity Rock*	5
20	Discussion: *A Charity Rock*	25
	"How Many of Each Kind?" Revisited (continued)	20
	Homework: *Back on the Trail*	5
21	*"How Many of Each Kind?" Revisited* (continued)	20

	Discussion: *Back on the Trail*		15
	Big State U		15
	Homework: *Inventing Problems*		0
22	Discussion: *Inventing Problems*		25
	Big State U (continued)		15

Creating Problems

	Homework: *Ideas for Linear Programming Problems*		10
23	Group exploration: *Ideas for Linear Programming Problems*		15
	Presentations: *Big State U*		15
	Producing Programming Problems		15
	Homework: *Beginning Portfolio Selection*		5
24	Discussion: *Beginning Portfolio Selection*		5
	Presentations: *POW 6: Shuttling Around*		20
	Producing Programming Problems (continued)		25
	Homework: *Just for Curiosity's Sake*		0
25	Discussion: *Just for Curiosity's Sake*		15
	Producing Programming Problems (continued)		30
	Homework: *"Producing Programming Problems" Write-up*		5
26	Presentations: *Producing Programming Problems*		45
	Homework: *Continued Portfolio Selection*		5

27	Presentations (continued): *Producing Programming Problems*	40
	Homework: "*Cookies*" *Portfolio*	10
28	*In-Class Assessment*	50
	Homework: *Take-Home Assessment*	0
29	Exam Discussion	30
	Unit Reflection	20

90-Minute Pacing Guide (18 days)

Day	Activity	In-Class Time Estimate
Cookies and Inequalities		
1	*How Many of Each Kind?*	45
	A Simpler Cookie	40
	Homework: *Investigating Inequalities*	5
2	Discussion: *Investigating Inequalities*	20
	How Many of Each Kind? (continued)	35
	My Simplest Inequality	30
	Homework: *Simplifying Cookies*	5
3	Discussion: *Simplifying Cookies*	15
Picturing Cookies		
	Picturing Cookies—Part I	50
	Introduce: *POW 4: A Hat of a Different Color*	15
	Homework: *Inequality Stories* and *Healthy Animals*	10
4	Discussion: *Inequality Stories*	10
	Discussion: *Healthy Animals*	10
	Picturing Cookies—Part II	25
	What's My Inequality?	45
	Homework: *POW 4: A Hat of a Different Color*	0
5	*Feasible Diets*	20
	Picturing Pictures (continued)	45
Using the Feasible Region		

	Profitable Pictures	25
	Homework: *Curtis and Hassan Make Choices*	0
6	Discussion: *Curtis and Hassan Make Choices*	15
	Profitable Pictures (continued)	30
	Finding Linear Graphs	30
	Presentations: *POW 4: A Hat of a Different Color*	15
	Homework: *You Are What You Eat*	0
7	Discussion: *You Are What You Eat*	20
	Hassan's a Hit	40
	Introduce: *POW 5: Kick It!*	25
	Homework: *Changing What You Eat*	5
8	Discussion: *Changing What You Eat*	30
	Rock 'n' Rap	35
	A Rock 'n' Rap Variation	25
	Homework: *Getting on Good Terms*	0
9	Discussion: *Getting on Good Terms*	15
	Rock 'n' Rap on the graphing calculator	20
	Going Out for Lunch	30
Points of Intersection		
	Get the Point	25
	Homework: *Only One Variable*	0
10	Discussion: *Only One Variable*	35
	Get the Point (continued, with presentations)	45
	Homework: *Set It Up*	10

11	Discussion: *Set It Up*	25
	A Reflection on Money	35
	Presentations*: POW 5: Kick It!*	20
	Introduce: *POW 6: Shuttling Around*	10
	Homework: *More Linear Systems*	0
12	Discussion: *More Linear Systems*	30
Cookies and the University		
	"How Many of Each Kind?" Revisited	50
	Homework: *A Charity Rock*	10
13	Discussion: *A Charity Rock*	25
	"How Many of Each Kind?" Revisited (continued)	20
	Back on the Trail	40
	Homework: *Inventing Problems*	5
14	Discussion: *Inventing Problems*	30
	Big State U	50
Creating Problems		
	Homework: *Ideas for Linear Programming Problems*	10
15	Group exploration: *Ideas for Linear Programming Problems*	15
	Presentations: *POW 6: Shuttling Around*	20
	Producing Programming Problems	45
	Homework: *Beginning Portfolio Selection* and *Just for Curiosity's Sake*	10

16	Discussion: *Beginning Portfolio Selection*	5
	Discussion: *Just for Curiosity's Sake*	10
	Producing Programming Problems (continued)	30
	Presentations: *Producing Programming Problems*	40
	Homework: "*Producing Programming Problems*" *Write-up, Continued Portfolio Selection,* and "*Cookies*" *Portfolio*	5
17	Presentations (continued): *Producing Programming Problems*	40
	In-Class Assessment	50
	Homework: *Take-Home Assessment*	0
18	Exam Discussion	30
	Unit Reflection	20

Materials and Supplies

All IMP classrooms should have a set of standard supplies and equipment, and students are expected to have materials available for working at home on assignments and at school for classroom work. Lists of these standard supplies are included in the section "Materials and Supplies for the IMP Classroom" in *A Guide to IMP.* There is also a comprehensive list of materials for all units in Year 2.

Listed here are the supplies needed for this unit. General and activity-specific blackline masters are available for presentations on the overhead projector or for student worksheets. The masters are found in the *Cookies* Unit Resources under *Blackline Masters*.

Cookies

- Graph paper transparencies (several per group)
- Colored pencils

More About Supplies

- Graph paper is a standard supply for IMP classrooms. Blackline masters of 1-Centimeter Graph Paper, 1/4-Inch Graph Paper, and 1-Inch Graph Paper are provided so you can make copies and transparencies. (You'll find links to these masters in "Materials and Supplies for Year 2" of the Year 2 guide and in the Unit Resources of each unit.)

Assessing Progress

Cookies concludes with two formal unit assessments. In addition, there are many opportunities for more informal, ongoing assessment throughout the unit. For more information about assessment and grading, including general information about the end-of-unit assessments and how to use them, consult the *Year 2: A Guide to IMP* resource.

End-of-Unit Assessments

This unit concludes with in-class and take-home assessments. The in-class assessment is intentionally short so that time pressures will not affect student performance. Students may use graphing calculators and their notes from previous work when they take the assessments.

Ongoing Assessment

Assessment is a component in providing the best possible ongoing instructional program for students. Ongoing assessment includes the daily work of determining how well students understand key ideas and what level of achievement they have attained in acquiring key skills.

Students' written and oral work provides many opportunities for teachers to gather this information. Here are some recommendations of written assignments and oral presentations to monitor especially carefully because they will offer insight into student progress.

- *Inequality Stories,* Part I: This assignment will give you information about students' understanding of how real-life contexts can be expressed in algebraic terms using inequalities.

- *Profitable Pictures:* This activity will tell you how well students understand how profit lines can be used to determine an optimal value.

- *Changing What You Eat:* In this assignment, students will demonstrate their understanding of how changing specific parameters in a problem affects the solution.

- *Get the Point:* This investigation will give you insight into students' abilities to think about systems of linear equations in flexible ways.

- *A Reflection on Money:* This assignment will give you information about students' comfort levels with solving systems of linear equations.

- *"How Many of Each Kind?" Revisited:* This activity will tell you how well students have synthesized the ideas of the unit.

- *Producing Programming Problems:* This assignment will tell you how well students understand the components of a linear programming problem.

Supplemental Activities

Cookies contains a variety of activities at the end of the student pages that you can use to supplement the regular unit material. These activities fall roughly into two categories.

- **Reinforcements** increase students' understanding of and comfort with concepts, techniques, and methods that are discussed in class and are central to the unit.

- **Extensions** allow students to explore ideas beyond those presented in the unit, including generalizations and abstractions of ideas.

The supplemental activities are presented in the *Teacher's Guide* and the student book in the approximate sequence in which you might use them. Below are specific recommendations about how each activity might work within the unit. You may wish to use some of these activities, especially the later ones, after the unit is completed.

***Who Am I?* (reinforcement or extension):** This logic problem is a good follow-up to the discussion of *POW 4: A Hat of a Different Color*.

***Algebra Pictures* (extension):** You might use this activity after discussing *Picturing Cookies—Part II*. It extends ideas in the unit by including nonlinear as well as linear inequalities.

***Find My Region* (reinforcement):** You might use this activity, which provides a lighthearted setting in which students practice finding equations for straight-line graphs and inequalities for half planes, after discussing *What's My Inequality?*

***Kick It Harder!* (extension):** Students can work on this activity any time after concluding the discussion of *POW 5: Kick It!*

***More Cereal Variations* (extension):** This activity asks students to create some additional variations to the cereal activities, *You Are What You Eat* and *Changing What You Eat.* If time allows, you may want to use *More Cereal Variations* in class after the discussion of these activities.

***Rap Is Hot!* (reinforcement):** You can assign this variation on *Rock 'n' Rap* any time after the discussion of that activity, perhaps after discussion of *A Rock 'n' Rap Variation.* You may want to wait until after students have examined *Rock 'n' Rap* on the graphing calculator.

***How Low Can You Get?* (extension):** This activity is appropriate for use beginning late in *Using the Feasible Region,* after students have had some experience examining the effects of changing parameters in a problem, as in *Changing What You Eat* and *A Rock 'n' Rap Variation.*

***Shuttling Variations* (extension):** This activity presents two major generalizations of the problem in *POW 6: Shuttling Around.* You may want to assign some or all of this activity as part of the POW itself.

***And Then There Were Three* (extension):** You might have students work on this activity after *Get the Point* and after they have had experience making up two-equations/two-unknown problems in *Inventing Problems.*

***An Age-Old Algebra Problem* (extension):** This activity is a follow-up to the previous supplemental activity, *And Then There Were Three.*

Cookies and Inequalities

Intent

The activities in *Cookies and Inequalities* introduce the unit problem. In addition, they extend students' understanding of linear equations to the linear inequalities they will encounter throughout this unit.

Mathematics

In the unit problem, a bakery that makes and sells two kinds of cookies must decide—given limits on ingredients, oven space, time, and cost—how many of each kind to make in order to maximize profit. In *Cookies and Inequalities*, students write these limiting conditions, or **constraints**, as a system of **linear inequalities**. They then look for specific numbers that satisfy the full set of constraints and determine the profit in each case.

Linear inequalities can be manipulated, using a set of rules, into **equivalent inequalities**. In these activities, students encounter the one difference between these rules and the rules for manipulating equations, first encountered in *The Overland Trail.*

Progression

Students begin *Cookies and Inequalities* by examining the unit problem. They then have opportunities to extend their symbol-manipulation skills from equations to inequalities.

How Many of Each Kind?

A Simpler Cookie

Investigating Inequalities

My Simplest Inequality

Simplifying Cookies

How Many of Each Kind?

Intent

Students are introduced to the unit problem and explore the situation informally.

Mathematics

In the unit problem, students deal with a set of constraints on ingredients, oven space, time, and cost as they try to maximize the profit from sales of two kinds of cookies. Each of the constraints can be expressed as a linear inequality in two variables, and the profit is a linear function of those same two variables. In this activity, students look for specific numeric examples that fit the constraints, and then they determine the profit for each example. They do this work without a formal introduction to linear programming, instead using their prior knowledge to follow the constraints introduced in the story.

Progression

Students work in groups to explore the central unit problem. They post their findings for later reference. In a class discussion, they develop inequalities to represent the constraints and a symbolic expression for the profit.

Approximate Time

70 minutes

Classroom Organization

Groups and whole class

Doing the Activity

Have students read the activity, perhaps allowing several students to read portions of it aloud. Then have students work in groups on the questions.

Part of the challenge of this problem is keeping track of all the numbers. Let students develop their own ways of organizing the information; one of the goals of the unit is to find ways to improve on their initial methods.

As you observe, make sure students realize that the numbers they are looking for are in *dozens*—for example, "4 dozen plain, 3 dozen iced," not "48 plain, 36 iced."

Students may at first think they have to make use of all the ingredients, oven space, and preparation time available. Let them become aware on their own that this is not required and is, in fact, impossible. You might ask such questions as, Could the Woos make 1 dozen of each kind? What about 3 dozen plain and 5 dozen iced?

Discussing and Debriefing the Activity

To prepare for a discussion, begin a chart like the following one that includes group number or name, dozens of plain cookies, dozens of iced cookies, dough used, icing used, time used, oven space used, and a profit column for plain cookies sold, iced cookies sold, and total profit. As groups find combinations for which they have enough ingredients, they can add their results to the chart even if they duplicate another group's combination. If that happens, ask the group to find and post an additional combination that works.

Group	Plain cookies (dozen)	Iced cookies (dozen)	Dough used	Icing used	Time used	Oven space used	Profit: plain cookies	Profit: iced cookies	Total profit

Have group representatives present their organizational schemes for keeping track of and computing the profit for various combinations of cookies. Then have representatives of other groups offer other possible combinations.

As combinations are suggested, ask students to check whether they satisfy the conditions by calculating the dough, icing, oven space, and preparation time required for that combination and determining whether the results fit the conditions. Are you sure that this combination fits all the conditions? How do you know?

Introduce the term **constraint** as a synonym for *condition.* You might remind students that they encountered this term when they were forming families in the Year 1 unit *The Overland Trail.*

Ask students how and when to compute the profit for each combination. They may recognize that it makes sense to wait until they can establish whether a combination fits all the constraints before they make the calculation.

It is just as important for students to understand why a combination is excluded as it is to show that it is included. They must also recognize that each condition is a separate constraint and that a combination must satisfy all four constraints. Ask, What is a combination that does *not* fit all the constraints? Which constraint or constraints does it fail to satisfy?

Constraints as Inequalities

Choose one of the conditions, such as the amount of dough available, to focus on and ask, How did you decide whether a combination fits this constraint? As a class, develop a statement that tells whether a combination fits that constraint; for example, *Take the number of dozens of iced cookies, multiply that by 0.7, and add the number of dozens of plain cookies. The result cannot be more than 110.*

You may have to start with a less-detailed statement and gradually get students to refine it. Questions such as **Do we want to multiply by the number of cookies?** can help prompt refinements such as talking about the number of *dozens* of cookies.

How could you express the "cookie dough constraint" symbolically? Suggest that students choose variables to represent "number of dozens of plain cookies" and "number of dozens of iced cookies." Using P and I for these two variables, for example, they should realize that the dough constraint can be expressed by the inequality $P + 0.7I \leq 110$.

Then have students work in their groups to write verbal as well as symbolic statements for the other constraints. It is important that everyone be able to deal with both ways of expressing each condition. You may want to randomly ask groups to present verbal or symbolic expressions. The class should end up with a set of constraint inequalities that looks something like this:

$P + 0.7I \leq 110$	(amount of cookie dough)
$0.4I \leq 32$	(amount of icing)
$P + I \leq 140$	(amount of oven space)
$0.1P + 0.15I \leq 15$	(amount of preparation time)

Have students record these constraint inequalities on chart paper. Post the chart for later use, perhaps titling it "Cookie Constraints."

Have groups use the inequalities to check that the combinations discussed earlier really do satisfy the constraints. In doing so, students will be repeating their earlier computations. But they should go through the process at least once to confirm that the symbolic inequalities are saying the same thing as the verbal expressions of the conditions.

The Profit Expression

Now ask groups to develop a symbolic expression for the profit. If they have difficulty, have them examine the chart showing profit computed for various combinations. They should get the expression $1.5P + 2I$.

Students often make the mistake of treating the profit expression as another constraint. You may want to ask, **Why isn't the profit expression included in our list of constraints?**

Restrictions on P and I

Because of the problem context, P and I must be whole numbers. Discuss this important issue now if a student introduces it. Otherwise, you might ignore it for now, as the issue will be more engaging and meaningful for students if it arises initially with them or in the natural context of discussions of graphs or feasible regions later in the unit.

When the issue does come up, note that it has two aspects:

- Neither P nor I can be negative.

- P and I must be integers. (Allowing P and I to be multiples of $\frac{1}{12}$ would also make sense.)

If students raise either of these issues, you can broaden the discussion to include both, perhaps asking whether there are other restrictions on the "eligible" values for P and I. You might point out that the question of eligibility depends on the problem situation; negative or noninteger solutions make sense in some problems but not in others.

Concerning the restriction to nonnegative numbers, you might ask students if they can make this restriction by writing additional constraints—specifically, by expressing the condition of "not being negative" as inequalities. They should notice that adding the inequalities $P \geq 0$ and $I \geq 0$ to the constraints list fixes this aspect of the model. Write these inequalities on the posted constraint chart.

The issue of avoiding noninteger values is more complex, as it cannot be handled by additional inequalities. Perhaps the best approach is simply to explain that there is no easy way to handle it and that students should ignore it for now. Emphasize that this means they will need to be especially careful later on to check whether their solutions make sense. If it turns out that the solution that provides the maximum profit is somehow "ineligible," students will have to decide where to go from there. (It turns out in this problem that the combination with the maximum profit does have whole-number values for P and I.)

You can also take this opportunity to talk about mathematical modeling. Point out that we often need to simplify or ignore certain aspects of a problem in order to get a usable mathematical description. Reintroduce the term **mathematical model** for an abstract description of a real-world situation.

Finally, bring out that when simplifications are made, it becomes especially important to refer back to the original conditions of the problem after the mathematical analysis is completed.

Key Questions

Could the Woos make 1 dozen of each kind? What about 3 dozen plain and 5 dozen iced?

Are you sure that this combination fits all the conditions? How do you know?

What is a combination that does *not* fit all the constraints? Which constraint or constraints does it fail to satisfy?

How did you decide whether a combination fits this constraint?

How can you express this constraint symbolically?

Why isn't the profit expression included in our list of constraints?

A Simpler Cookie

Intent

By examining a simpler version of the unit problem, students begin to build an understanding of the relationship between the problem constraints and the resulting profit.

Mathematics

Working through a simpler problem is a valuable problem-solving strategy for complicated questions. In this activity, students will find numeric values for the numbers of plain and iced cookies that fit the constraint on preparation time:

$$0.1P + 0.15I \leq 15$$

They will compute the resulting profit for each pair of values, where the profit is a function of P and I:

$$Profit = 1.5P + 2I$$

Having only one constraint makes this problem much simpler than the original.

Progression

Students work on the activity individually and share results in groups and with the class.

Approximate Time

25 minutes for activity (at home or in class)

15 minutes for discussion

Classroom Organization

Individuals, then groups, followed by whole-class discussion

Doing the Activity

You may want to have volunteers read the activity, which presents a simpler version of the unit problem, aloud.

Discussing and Debriefing the Activity

Have students discuss their solutions briefly in groups. Then ask representatives from one or two groups to present their group's solution.

Students do not need to find a definitive resolution to this problem. They don't need to be able to prove, or even be sure, that a particular choice is optimal. They will be learning how to deal with problems of this type over the course of the unit.

Some students will realize that to maximize profit, the Woos should maximize the number of plain cookies. In other words, they should make 150 dozen plain cookies and no iced cookies. They may arrive at this conclusion in various ways, including guess-and-check. Specifically, in 15 hours, the Woos can make 150 dozen plain cookies, which will yield a profit of $225. By comparison, in 15 hours, they can make 100 dozen iced cookies, which will yield a profit of only $200.

Investigating Inequalities

Intent

In this activity, students investigate what manipulations of an inequality will preserve its truth.

Mathematics

Beginning in the Year 1 unit *The Overland Trail,* students developed procedures for changing equations into equivalent forms, often for the purpose of solving them. Now they will begin to determine procedures that change inequalities—like those that arise throughout this unit—into equivalent forms. In this activity, they will start with numeric inequalities like $-2 \leq 7$ and explore what happens to the truth of these statements when they add, subtract, multiply, or divide both sides by the same number. They will also confront the specific case of multiplying or dividing both sides of an inequality by the same negative number. In addition, they will use one-dimensional linear graphs to represent and justify their results.

Progression

Students work on this activity individually and then compare their results in groups and in a class discussion.

Approximate Time

5 minutes for introduction

25 minutes for activity (at home or in class)

20 minutes for discussion

Classroom Organization

Individuals, followed by group discussion and whole-class discussion

Doing the Activity

For this work, students need to know how to use the terms and symbols for inequalities for both positive and negative numbers. For example, they should recognize that -8 is less than -6 and that the statement $5 \geq -7$ is true.

If they need a review of these ideas and of inequality notation, you will probably want to use the number line as the basic reference point for this discussion.

Discussing and Debriefing the Activity

If you review Part II of this activity first, number lines can then be used to explain the rules for preserving the truth of inequalities, which are explored in Part I. Ask volunteers to present their answers to Questions 4 through 7.

Students may have described the graph in Question 7 using two inequalities, $x > -1$ and $x \leq 3$. If so, explain that we usually combine such pairs into a single inequality, written as $-1 < x \leq 3$. (Rather than getting bogged down in distinctions between "and" and "or," you might suggest that students simply read the inequality $-1 < x \leq 3$ as "x is between −1 and 3, not including −1 and including 3.")

Part I: Manipulating Inequalities

In groups, have students share the conclusions they reached in Part I. Then conduct a class discussion to develop a list of manipulations that can be applied to a true inequality to produce other true statements.

The operations of multiplication and division may be more difficult than addition and subtraction. For example, students might overlook the case of multiplying or dividing by a negative number or simply say that the resulting inequality is false.

Students' eventual lists should state that if they start with a true inequality, any of the operations presented here will produce a true statement.

- Add the same number to both sides of the inequality.
- Subtract the same number from both sides of the inequality.
- Multiply both sides of the inequality by the same positive number.
- Divide both sides of the inequality by the same positive number.
- Multiply both sides of the inequality by the same negative number and reverse the direction of the inequality sign.
- Divide both sides of the inequality by the same negative number and reverse the direction of the inequality sign.

Post the class list once students have completed it.

Multiplying or Dividing by Negative Numbers

The rules for working with inequalities differ from those for equations when multiplying or dividing by a negative number.

To raise this potentially confusing topic, you might ask, **What happens if you multiply both sides of 4 > 3 by −2?** If students have followed the instructions to consider both positive and negative numbers, they will have encountered this complication. They may simply state that multiplying by a negative number gives a false statement. For example, they may say that $4(-2) > 3(-2)$ is not true, because −8 is not more than −6. Ask whether there is any way to adjust the statement $4(-2) > 3(-2)$ to make it true. You may have to suggest changing the inequality sign. Using examples as needed, lead students to realize that they will get a true statement if they reverse the direction of the inequality sign.

This reversal, a departure from what happens with equations, is important for students to understand. Ask for their explanations. **Why does multiplying by a negative number reverse the inequality?**

One approach to explaining the reversal of inequalities is based on the number line. Begin with an inequality such as 8 > 5, and ask students to explain this statement in terms of the number line. They should simply point out that 8 is to the right of 5.

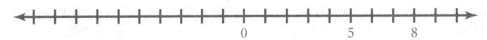

Ask for a volunteer to show what adding the same thing to both sides of the inequality does. Students should recognize that it shifts the two numbers, 8 and 5, equal distances—to the right if the addend is positive, to the left if the addend is negative. For example, they might use the diagram to illustrate adding 2 to both sides and then to show that the resulting inequality, 10 > 7, is also true.

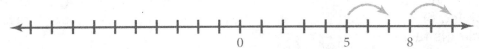

Then ask students to multiply both sides of the original inequality, 8 > 5, by 2. They will notice that both numbers move to the right, with 8 • 2 ending up to the right of 5 • 2, preserving the relationship.

Next, have students multiply each side of the inequality by −1 and describe what happens in terms of the number line. They might use a phrase like "on the other side of 0" to describe where −5 and −8 are found.

Bring out the symmetry of the situation around 0, focusing on the idea that because 8 is farther to the right than 5, its reflection around 0 ends up farther to the left.

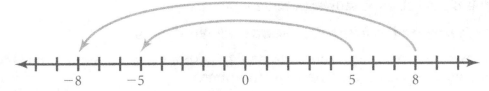

Key Questions

What manipulations of an inequality will preserve its truth?

What happens if you multiply both sides of 4 > 3 by −2?

What happens if you multiply both sides of an inequality by a negative number?

Why does multiplying by a negative number reverse the inequality?

If you have an inequality that is not true, is there any way to adjust the inequality sign to make it be true?

My Simplest Inequality

Intent

Students continue their exploration of the symbolic manipulation of inequalities.

Mathematics

This activity emphasizes the notion of an **equivalent inequality** in one or two variables. Students establish equivalence by confirming that specific examples that make the original inequality true also make the new inequality true. They also check whether examples that do not work in the original inequality also do not work in the new one. They use these procedures to solve one-variable inequalities and then to simplify two-variable inequalities.

Progression

Students work on the two parts of this activity in groups and share their results in a class discussion.

Approximate Time

25 minutes

Classroom Organization

Groups, followed by whole-class discussion

Doing the Activity

Remind students that, in the Year 1 unit *The Overland Trail,* they applied principles similar to those just discussed in *Investigating Inequalities* in order to replace an equation with a simpler one having the same solution or solutions. Review the concept of **equivalent equations**, and tell students that a similar concept applies to inequalities.

You might begin with a simple inequality such as $x + 3 < 9$. Ask students what they can do to get an inequality with the same solutions. They will likely suggest subtracting 3 from both sides to get $x < 6$. They can test this by finding numbers that fit either $x + 3 < 9$ or $x < 6$ and verifying that they fit the other inequality as well.

Bring out that the goal for inequalities with one variable is to get a statement that precisely describes the numbers that satisfy the inequality. For example, the fact that $x < 6$ is equivalent to $x + 3 < 9$ shows that the solutions to $x + 3 < 9$ are all numbers less than 6.

Now give students an inequality with two variables, such as $0.5x \geq 2y - 7$, and ask for a simpler inequality that is equivalent. For instance, they might multiply both sides by 2 to get $x \geq 4y - 14$ and then add 14 to both sides to get $x + 14 \geq 4y$.

Bring out that there is no single best form for this inequality because, unlike one-variable inequalities, no equivalent inequality shows the solution. For inequalities with two or more variables, the goal of using equivalent inequalities is to find a form that is comparatively easy to understand.

Let students work in groups on the activity. Circulate to confirm that they are finding the solutions in Part I and getting a variety of equivalent forms in Part II.

Discussing and Debriefing the Activity

As students share their work for Question 1, bring out that there is often more than one way to simplify a given inequality. For example, here are two sequences of steps for simplifying $4 - 2x > 7 + x$ (Question 1d).

$4 - 2x > 7 + x$	$4 - 2x > 7 + x$
$-2x > 3 + x$	$4 > 7 + 3x$
$-3x > 3$	$-3 > 3x$
$x < -1$	$-1 > x$

Some students may prefer to keep the variable on the left side of the inequality, as in the first sequence. Others may prefer to have a positive coefficient for x, as in the second sequence, and avoid the issue of reversing the inequality. In either case, be sure they understand that the results, $x < -1$ and $-1 > x$, are equivalent.

For Question 2, you might have one or two students present their examples for Questions 2a and 2c and show whether they do or do not fit the original inequality as required. Be sure students recognize the distinction between knowing that the two inequalities have solutions in common and actually proving their equivalence. Bring out that it would be impossible to verify case by case that every pair that fits either inequality also fits the other.

Then go over Question 3. As with Question 1, bring out that there is more than one possible sequence of steps.

Key Questions

What do you call two equations with the same solution or solutions?

How can you write the inequality in simplest form?

Simplifying Cookies

Intent

In this final activity in *Cookies and Inequalities,* students practice finding equivalent inequalities, this time using the unit problem's constraints.

Mathematics

Throughout the unit, students will look at many combinations of cookies that the Woos might bake. By creating equivalent inequalities, they can make this work simpler. Students will also be graphing inequalities, which will be easier if they can flexibly find equivalent inequalities. The activity highlights the one issue that makes manipulating inequalities different from manipulating equations: the effect of multiplying or dividing by a negative number.

Progression

Students work on the activity individually and share results in a class discussion.

Approximate Time

5 minutes for introduction

15 minutes for activity (at home or in class)

15 minutes for discussion

Classroom Organization

Individuals, followed by whole-class discussion

Doing the Activity

The inequalities presented in this activity have likely already been posted in the classroom. Suggest to students that these inequalities can be written in easier forms. As there are many combinations of plain and iced cookies that the Woos can choose to bake, writing equivalent inequalities may help students quickly develop a long list of combinations.

Discussing and Debriefing the Activity

Have students present equivalent inequalities for each of the given cookie inequalities. Try to elicit a variety of responses. For example, students might multiply both sides of the preparation-time constraint, $0.1P + 0.15I \leq 15$, by 100 to obtain

$$10P + 15I \leq 1500$$

They may observe that this inequality is also equivalent to $2P + 3I \leq 300$. There is no "best" equivalent inequality and no need to have students look for the greatest common divisor of the coefficients. It is often helpful, though, to eliminate decimals and to simplify inequalities when it is easy to do so.

Students should notice that, because the icing inequality, $0.4I \leq 32$, involves only one variable, it does have a simpler equivalent, $I \leq 80$. The oven-space inequality, $P + I \leq 140$, is likely already in simplest form.

You may want to skip discussion of Question 2, as students will be doing the same kind of work in the next activity.

Key Questions

Why is it helpful to look at equivalent inequalities for the unit problem's constraints?

What are the advantages of working with integer coefficients?

What is the advantage of solving for one variable in terms of the other?

Picturing Cookies

Intent

The activities in *Picturing Cookies* build on students' developing abilities to write inequalities to express the constraints in a linear programming context. They also introduce students to the graphical representation of systems of inequalities.

Mathematics

The graph of the solutions for an inequality in two variables is a region known as a **half plane**. The boundary of that region is the inequality's associated equation. The graphical solution of a system of such inequalities, known as the **feasible region** for the system, is the intersection of the inequalities' half planes. The feasible region for the unit problem, which will be created by students in *Picturing Cookies,* is shown in the graph here. Students will also find the feasible regions for three other linear programming contexts.

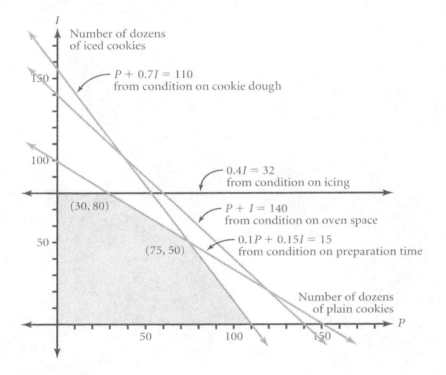

**The Feasible Region
for the Cookie Problem**

Progression

The activities first introduce students to the idea of graphing an inequality. Later in the activities, students will place the graphs of all inequalities for a given context on one set of axes to define the feasible region. In addition, students will begin work on the first POW of the unit.

Picturing Cookies—Part I

Inequality Stories

POW 4: A Hat of a Different Color

Healthy Animals

Picturing Cookies—Part II

What's My Inequality?

Feasible Diets

Picturing Pictures

Picturing Cookies—Part I

Intent

Students are introduced to the graphical representation of a linear inequality.

Mathematics

A linear inequality, such as the unit problem's oven-space constraint $P + I \leq 140$, divides the coordinate plane into two **half planes** with a boundary formed by the corresponding linear equation, in this case $P + I = 140$. One half plane contains ordered pairs that satisfy the inequality. In this activity, students build the half plane by testing and coloring points that either satisfy or do not satisfy the inequalities defined by the constraints in the unit problem. Graphing linear inequalities will soon lead students to the notion of a **feasible region**.

Progression

Students work in groups to test and plot many points on a common graph, using one color for points that fit the condition and another for points that fail to fit the condition. They compare their results with those of other groups.

Approximate Time

30 minutes for activity

20 minutes for discussion

Classroom Organization

Groups and whole class

Materials

1" grid chart paper

Graph paper transparencies and overhead pens

Doing the Activity

In this activity, students work from scratch to develop the graph of one or more of the inequalities from the unit problem. Students are often taught to graph an inequality by first graphing the corresponding equality and then shading the appropriate side of that line. However, having students follow the activity instructions, even if they already know this shortcut, is important. The experience of actually plotting points that fit the inequality and considering the graph as it "evolves" will afford students a much deeper appreciation of the relationship between the graph of an equation and the graph of an inequality.

To introduce the activity, refer the class to the list of constraint inequalities compiled for the unit problem, and ask, How might you obtain a geometric picture of these constraints?

Choosing Axes

One step in creating a graph is deciding on the axes. Talk with the class about the issue of choosing which axis will represent which variable. Bring out that there is no reason why a particular variable should be on one axis or the other, as neither variable is more "dependent" or "independent" than the other. Also bring out that it will be easier for students to understand and compare their graphs if the class comes to a consensus.

Graphs in this Teacher's Guide and in the take-home assessment for the unit use the horizontal axis for P and the vertical axis for I, so you may want to have students do the same. (In some activities, such as *Healthy Animals,* each student or group will make an individual choice of axes.)

The Meaning of the Graph of an Inequality

Ask, What does it mean to graph an inequality in two variables? Help students understand that, as with graphing an equation, graphing an inequality means marking all the number pairs that fit the condition. Students may not realize that the graph of an inequality is a two-dimensional area rather than a line or a curve; let them discover this through the activity.

Give each group two sheets of gridded chart paper for Question 1. One sheet is for students' initial experimentation; the other is for use after they have a sense of what their scales should be. Groups moving on to Question 2 will need more chart paper.

After each group member finishes checking a number pair, he or she should add that point, in the appropriate color, to the graph. The idea is to accumulate a lot of points quickly so that students see the graph emerge and eventually recognize the overall relationship.

If students are stuck, ask what color should be used for the number pair $P = 5, I = 10$. What color should (5, 10) be? Because this pair satisfies the inequality $P + I \le 140$, the point (5, 10) should be plotted with the first color.

Check that groups include examples that don't fit the inequality as well as those that do. For instance, the pair $P = 100, I = 100$ fails to satisfy the inequality, so the point (100, 100) should be plotted with the second color.

As you circulate, you may want to ask groups about a pair like (40, 100) to bring out that if a number pair fits the equation $P + I = 140$ then it also fits the inequality $P + I \le 140$.

Selecting Scales

After some experimentation, students may find that they can't fit points that don't work within their existing coordinate system. You might need to suggest that they redraw their graphs with new scales, first thinking about what scales might be appropriate. **What is the maximum value you will need on each axis?** They may realize that for this inequality, the largest value they will need for either variable for points that satisfy the inequality is 140. However, they should choose scales so that they can also show points that do not satisfy the inequality.

You may want to bring up the issue of scale selection in the class discussion later, asking whether any groups had to adjust their scales after starting on the problem.

The "Big Picture"

Encourage students to keep plotting points, both points that satisfy and those that do not satisfy the inequality, until they realize that the boundary between satisfying the inequality and not satisfying the inequality is the line corresponding to the equation $P + I = 140$. This is the "big picture" referred to in Question 1.

Question 2

If groups move on to Question 2, emphasize that they should treat each constraint as a separate problem on its own set of axes. (In *Picturing Cookies—Part II,* students will look at combining graphs of inequalities on a single set of axes.)

The inequality for icing may present special problems, as it involves only one variable. You might ask, **Are there any restrictions for the dozens of plain cookies when you are focused only on how much icing the Woos have?** This should help students determine several ordered pairs in which P can be any number while I has many values (technically, an infinite number of values) but also has an upper limit.

Discussing and Debriefing the Activity

Groups will probably grasp the main idea of this activity at different rates. As soon as everyone has finished Question 1, have groups present their diagrams to the class. Ask, **What do the different colors represent?** Make sure presenters relate the colors to the concepts of *satisfying the inequality* and *not satisfying the inequality*.

The main goal of this discussion is to articulate the "big picture" referred to in Question 1: the relationship between the graph of an inequality and the graph of the corresponding equation. Make sure students understand that the graph of the equation forms the boundary between points that satisfy the inequality and points that do not. In their statements, students should express this principle in their own words.

Ask whether someone can explain why all the solutions to the inequality are on one side of the graph of $P + I = 140$. **Why is $P + I = 140$ the boundary?** A student might say that if you move to the left or down from the line $P + I = 140$, either P or I will decrease, so the value of $P + I$ will go down. If this value is equal to 140 *on the line,* then it is *less than* 140 to the left of the line.

Introduce the term **half plane** to describe the set of points that fit the inequality. Make sure students realize that a plane is infinite in extent. They might describe a plane as "an infinite flat surface."

Once students clearly understand why the graph of the equation $P + I = 140$ forms the boundary between the two colors, review the concept of a **linear equation** by asking, **Why is the graph of $P + I = 140$ a straight line?** This need be only an intuitive discussion, perhaps based on the idea that each increase in the value of P must be matched by an equivalent decrease in the value of I.

Bring out that all of the algebraic expressions in the unit problem—in both the constraints and the profit expression—are linear. Explain that an inequality involving linear expressions is called a **linear inequality**.

Have groups that worked with other constraints, in Question 2, present their results. Take some time discussing the constraint $0.4I \leq 32$, because of the absence of the variable P. You may want to suggest rewriting this inequality as $0P + 0.4I \leq 32$. You can also refer back to the problem context, asking, **If the only restriction on the Woos was the amount of icing they had, how many dozens of plain and of iced cookies could they make?** Students should realize that this constraint by itself does not limit the number of dozens of plain cookies at all.

Key Questions

How might you obtain a geometric picture of these constraints?

What does it mean to graph an inequality in two variables?

What color should (5, 10) be? What color should (40, 100) be?

What is the maximum value you will need on each axis?

Are there any restrictions for the dozens of plain cookies when you are focused only on how much icing the Woos have?

What do the different colors represent?

Why is $P + I = 140$ the boundary?

Why is the graph of $P + I = 140$ a straight line?

If the only restriction on the Woos was the amount of icing they had, how many dozens of plain and of iced cookies could they make?

Inequality Stories

Intent

This activity offers students more experience relating inequalities to real-world situations.

Mathematics

The IMP curriculum, beginning in *The Overland Trail,* has stressed the representational nature of symbolic algebra. Students have written expressions and equations that summarize patterns and relationships in real-world contexts. When these expressions and equations have been manipulated, the manipulations have been connected to these contexts. Further, the curriculum has fostered algebraic thinking by asking students to create contexts from expressions and equations. In this activity, students extend this work to inequalities.

Progression

Students work on the activity individually and share results in a class discussion.

Approximate Time

5 minutes for introduction

20 minutes for activity (at home or in class)

15 minutes for discussion

Classroom Organization

Individuals, then groups, followed by whole-class discussion

Doing the Activity

When you introduce the activity, remind students that clear definitions of their variables will make their work easier. Also, if they first describe each situation in Part I in words, they will find the transition to writing an inequality smoother.

Discussing and Debriefing the Activity

Have students share their work in groups. You might assign one question to each group to discuss in detail. Any group working on Questions 3, 4, or 5 can select a couple of possible situations to report to the class. Remind students to make sure they have clearly defined their variables.

Question 1 should be straightforward, with an inequality like $LW \geq 150$ coming out of the presentation.

For Question 2, students should come up with something equivalent to $A + 2A + B + B < 200$. They might substitute numeric values to confirm that the inequality is correct.

For Questions 3 through 5, students will probably suggest a variety of situations. The responses for Question 5, for example, might involve the Pythagorean theorem or area.

POW 4: A Hat of a Different Color

Intent

In this POW, students work through an interesting and challenging logic puzzle.

Mathematics

This POW's focus is on logical thinking and developing convincing arguments. Students must organize the given information and devise a chain of reasoning that leads to a convincing answer.

Progression

This activity is introduced in class, with enough time for students to sort out the problem being posed. Students might benefit from some class time to explore the problem in groups. Approximately one week later, students present their findings.

Approximate Time

10 minutes for introduction

1 to 3 hours for activity (at home)

15 minutes for presentations a week or so later

Classroom Organization

Whole class, then groups and individuals, followed by whole class for presentations

Doing the Activity

Have volunteers read the POW aloud. Then help students grasp the problem by having them act out a similar but simpler situation. Two such scenarios are given here, each using one red hat and two blue hats. You can use pieces of colored paper to represent the hats. At this point, you might have students act out only Scenario 1. In a couple of days, if they seem stumped, you might present Scenario 2.

You might have students work in groups on this POW, with the suggestion that they develop a skit or some other way to act out portions of their presentations.

Scenario 1

Show the three hats to the class. Then have two students close their eyes. Put the red hat on the head of one student and a blue hat on the head of the other student. Hide the second blue hat.

Now ask the student with the red hat, **Open your eyes and look at the other student's hat. Can you tell what color you have?** This student will see a blue hat. Because there is another blue hat and a red one, and because either can be on his or her head, this student should realize that he or she can't be sure what hat he or she is wearing.

Now ask the student with the blue hat the same question. This student will see a red hat and should be able to deduce that her or his own hat must be blue, because there is only one red hat.

Scenario 2

Have two students close their eyes. This time put blue hats on both students. Have one student open his or her eyes. The student will see a blue hat. Because there is both a red hat and a blue hat still unaccounted for, this student can't tell the color of her or his own hat.

Then ask the second student what he or she can deduce *without opening his or her eyes.* The key insight is that if the first student had seen a red hat, that student would have known that he or she had a blue hat. Therefore, the second student must be wearing a blue hat. If the second student doesn't articulate this reasoning, ask whether anyone can help out.

Discussing and Debriefing the Activity

Have three students or groups present their findings.

A key concept is that certain possibilities are eliminated by the fact that Arturo and Belicia cannot figure out the color of their own hats. For example, had Arturo seen two red hats, he would have known that his was blue because there were only two red hats. That tells both Belicia and Carletta that at least one of them is wearing a blue hat. Similar reasoning, based on the fact that Belicia didn't know what color hat she had, gives Carletta the answer.

Supplemental Activity

Who Am I? (reinforcement or extension) is a logic problem that makes a good follow-up to this POW.

Healthy Animals

Intent

In this activity, students derive constraints and create graphs to represent another real-world context.

Mathematics

Students derive a set of constraints and create graphs for a new context. Several of these constraints are of the form $ax + by \geq c$, so their half planes will be above (to the right) of their boundaries. In addition, as in many linear programming contexts, the natural domains of the variables in this problem (in other words, the values the variables can take on in the given context) are nonnegative. Students will write new inequalities to represent this.

Progression

Students work on this activity individually. A class discussion then will raise several important issues about finding the appropriate half plane for an inequality and adding implicit constraints.

Approximate Time

5 minutes for introduction

20 minutes for activity (at home or in class)

15 minutes for discussion

Classroom Organization

Individuals, then groups, followed by whole-class discussion

Doing the Activity

Before they start their individual work on this activity, help students choose variables for expressing their inequalities. The class need not agree on which variable to use on which axis. If students make their own choices about this, it will make for interesting discussion.

Discussing and Debriefing the Activity

Have students discuss their solutions in groups and try to understand one another's work. Most will likely have put the amount of Food A on the horizontal axis. As you circulate, try to find some who put Food A on the vertical axis. Showing examples of both approaches will demonstrate the impact of choice of axes.

Before reviewing the specific questions, ask, **What differences do you notice in your work on this activity?** Students may mention that the variables have different names, that the axes are labeled differently, or that the scales vary.

Choosing Variables and Writing Inequalities

As a class, decide on variables and what they represent. **Let's all use the same variables. What variables shall we use? What do they represent?** The discussion that follows uses a and b for the number of ounces of Foods A and B, respectively, and puts a on the horizontal axis and b on the vertical axis.

Let representatives from several groups share the inequalities they developed. If necessary, have them restate their expressions using the variables decided on as a class. They will probably come up with these inequalities (or equivalents):

$$2a + 6b \geq 30 \qquad \text{(to guarantee enough protein)}$$

$$4a + 2b \geq 16 \qquad \text{(to guarantee enough fat)}$$

$$a + b \leq 12 \qquad \text{(to limit total intake)}$$

Graphs of the Inequalities

Once the inequalities are agreed on, let representatives from other groups present their graphs. If you found students who put Food A on the vertical axis, try to include one such presenter for each inequality.

Have students comment on the variations among the graphs for each inequality. Such variation may arise from different labeling of axes or different scales.

Because this situation will be revisited in *Feasible Diets,* it would be best to agree now on axis labels and scales. However, do point out that there is no "right" way to label or scale these graphs.

This is a good time to ask, **How can you tell which side of the line you want?** Clarify that the line itself is included as well, unless the inequality is strict.

Perhaps the simplest method is to pick a point off the line and determine whether it satisfies the inequality. Consider an easy point, such as $(0, 0)$. This point is below the line representing the equation $2a + 6b = 30$ and does not satisfy the inequality $2a + 6b \geq 30$. Therefore, the points that do satisfy the inequality form the half plane above the line.

Similarly, the point $(2, 0)$ is below the line $a + b = 12$ and satisfies the inequality $a + b \leq 12$. Therefore, the graph of $a + b \leq 12$ must be the half plane below the line $a + b = 12$. (It may seem obvious that $\geq$ means above and $\leq$ means below. But the inequality $a + b \leq 12$ is the same as $12 \geq a + b$, so the direction of the inequality sign doesn't by itself determine which side of the line represents the graph of the inequality. Also, it is not obvious what to do when there are variables on both sides of an inequality, as in $2x \geq y - 3$.)

First Quadrant Only

If it hasn't yet been discussed, this is a good time to bring up the fact that it often doesn't make sense in a real-world problem for variables to be negative. In *Healthy Animals,* both *a* and *b* must be nonnegative. It may be especially helpful to raise this idea now, because you can do so without limiting the values of the variables to integers. In *Healthy Animals,* all nonnegative numbers make sense.

While looking at the graph of one of the inequalities, such as $2a + 6b \geq 30$, ask whether a specific point is a possible choice. **Is (−1, 11) a possible choice for Curtis's pet's diet? Does it fit the constraints?** If students say no, argue that the coordinates satisfy the inequality—and in fact satisfy all three inequalities—and that therefore the point should be an option.

Students should say this doesn't make sense, as an amount of food cannot be negative. Emphasize that this condition does not come from any of the three constraint inequalities but instead from the meaning of the variables in the problem context.

You can suggest that students need to distinguish between the issue of finding possible diets for Curtis's pet and the issue of graphing the particular inequality. If they are just focusing on the latter, the point (−1, 11) should be included, because it satisfies the inequality. However, for the former, "−1 ounces of Food A, 11 ounces of Food B" is not an acceptable answer.

Suggest that they can take care of this issue by writing additional inequalities. **How can you impose this nonnegativity condition on the problem?** Help them come to the understanding that they could accomplish this by adding two additional constraints to the problem, $a \geq 0$ and $b \geq 0$.

Key Questions

What differences do you notice in your work?

What variables shall we use? What do they represent?

How can you tell which side of the line you want?

Is (−1, 11) a possible choice for Curtis's pet's diet? Does it fit the constraints?

How can you impose this nonnegativity condition on the problem?

Picturing Cookies—Part II

Intent

In this activity, students construct the feasible region for the unit problem by coloring, one at a time, the set of points that fits each constraint.

Mathematics

The set of inequalities that represent the constraints for the unit problem can be graphed on a single set of axes. The region of the plane that is the intersection of the half planes for the individual constraints is the **feasible region** for the problem. All of the points in the feasible region fit all of the constraints simultaneously.

Progression

Students work in groups to create a single graph of the unit problem's feasible region. You will probably want all students to make their own graphs so they can refer to them later and include them in their portfolios.

Approximate Time

30 minutes

Classroom Organization

Groups, followed by whole-class discussion

Materials

Grid chart paper

Doing the Activity

The major new idea introduced in this activity is the principle that the set of points that simultaneously satisfy more than one linear inequality is the *intersection* or *overlap* of the individual half planes for the separate inequalities.

As mentioned earlier, the values for *P* and *I* in the answer to the unit problem should probably be whole numbers, or at least multiples of $\frac{1}{12}$. If this issue hasn't come up before, introduce it in connection with this activity. You can raise it with individual groups as they work or wait for the class discussion to give students a chance to bring it up. To introduce this issue yourself, simply name a point that fits all of the constraints, such as (52.3, 28.7), and ask, **Does (52.3, 28.7) fit the constraints? Could these be the numbers of dozens of each kind of cookie?** Students should recognize that this point doesn't make sense in the context of numbers of dozens.

Suggest that groups start with complete graphs of the inequalities, including points whose coordinates are not whole numbers. Later on, they can consider how fully the inequalities model the real-world situation. If it turns out that the point that gives the maximum profit has coordinates that are "ineligible," students will have to decide where to go from there. (In this problem, the maximum profit does occur at a point with whole-number coordinates. In problems for which this is not the case, the task of finding the optimal whole-number solution can be quite complicated.)

Discussing and Debriefing the Activity

Ask for one or two volunteers to present the combined graphs from their groups. Encourage them to explain how they found their region, going through the inequalities one at a time. The rest of the class can ask questions like those listed here.

Where did that inequality come from?

What does that line represent?

Why do we want the points on this side of the line rather than on the other side?

Are points on the line part of the region?

What does the final graph tell you about the cookie problem?

Presenters should articulate that the points that have been colored every time represent the Woos' choices. Be sure that the "first quadrant only" issue has been dealt with by including the constraints $P \geq 0$ and $I \geq 0$.

Feasible Region and Feasible Points

Play up the importance of the final graph (Question 4), bringing out that it combines much of the information from the problem into a single picture. **How does seeing this combined graph compare to reading the verbal description of the situation?** Have students look back at the original problem, *How Many of Each Kind?*, to recognize what an achievement it is to represent so much information so simply.

Explain that in standard mathematics terminology this set of points is called the **feasible region** for the set of inequalities and that points in the region are called *feasible points*. You may want to comment on the everyday use of the word *feasible* to mean "possible" or "capable of being done."

The shaded area in the graph represents the feasible region for the cookie problem.

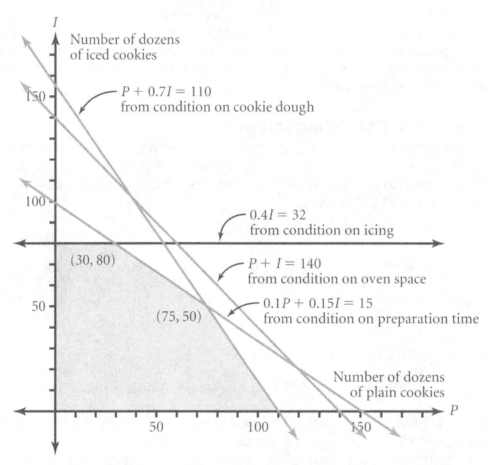

The Feasible Region
for the Cookie Problem

I

Number of dozens
of iced cookies

$P + 0.7I = 110$
from condition on cookie dough

$0.4I = 32$
from condition on icing

$P + I = 140$
from condition on oven space

$0.1P + 0.15I = 15$
from condition on preparation time

(30, 80)

(75, 50)

Number of dozens
of plain cookies

P

An Extraneous Constraint

Notice that the line $P + I = 140$, which comes from the oven-space limitation, misses the region completely. Ask students, **What does it mean that the oven-space line misses the feasible region?**

One way of expressing the significance of this fact is that, due to the other constraints, it wouldn't help the Woos to have an unlimited supply of oven space. Another way of looking at this is that because the Woos must satisfy all of the other constraints, they can't make full use of their available oven space.

Key Questions

Does (52.3, 28.7) fit the constraints? Could these be the numbers of dozens of each kind of cookie?

Where did that inequality come from?

What does that line represent?

Why do we want the points on this side of the line rather than on the other side?

Are points on the line part of the region?

What does the final graph tell you about the cookie problem?

How does seeing this combined graph compare to reading the verbal description of the situation?

What does it mean that the oven-space line misses the feasible region?

Supplemental Activity

Algebra Pictures (extension) challenges students to create graphs for linear and nonlinear inequalities.

What's My Inequality?

Intent

As in *Inequality Stories,* this activity emphasizes the connections among representations of relationships and continues to foster algebraic thinking.

Mathematics

Doing the opposite of what they have been doing so far in this unit, students will start with graphical representations of equations and inequalities and develop the corresponding algebraic statements. In the process, they will focus on the role of the equation as a boundary, the distinction between strict and nonstrict inequalities, and determining the direction of an inequality.

Progression

Students work on the activity individually, then share their work in groups and with the class.

Approximate Time

5 minutes for introduction

25 minutes for activity (at home or in class)

25 minutes for discussion

Classroom Organization

Individuals, then groups, followed by whole-class discussion

Materials

Graph paper or a copy of the activity's graphs

Doing the Activity

When you introduce the activity, you may want to mention that creating an In-Out table of points along the boundary line in the graphs of Part II can help students to identify the inequality represented.

Discussing and Debriefing the Activity

You might ask students to work in groups to prepare presentations of one of the questions. In Part I, make sure presenters describe how they developed their equations. Their methods will probably be rather informal, and the class will benefit from observing a variety of approaches. Whatever strategy a group presents, ask, **Does anyone have a different way to find the equation?**

Use Part II to review (1) the principle that the line itself forms the boundary of the graph of an inequality, (2) the distinction between strict and nonstrict inequalities, and (3) how to determine the direction of an inequality, perhaps by choosing a convenient point not on the boundary as a test case.

Key Question

Does anyone have a different way to find the equation?

Supplemental Activity

Find My Region (reinforcement) describes a game involving linear inequalities.

Feasible Diets

Intent

This activity is a follow-up to the introduction of feasible regions in *Picturing Cookies—Part II.* It uses the context presented in *Healthy Animals.*

Mathematics

Students find the intersection of the half planes they defined using inequalities in *Healthy Animals* to find the set of points that satisfy all the inequalities simultaneously. This feasible region is also defined by the implicit constraint that the solution is nonnegative.

Progression

Students work in groups to create their own graphs showing the feasible region for this context.

Approximate Time

20 minutes

Classroom Organization

Groups

Materials

Grid chart paper (optional)

Doing the Activity

Remind students that they graphed several individual inequalities in the activity *Healthy Animals.* Now they will put all their work together to find the "best" recommendations for Curtis's pet's diet.

You may want to ask that each group prepare a poster showing the feasible region.

Discussing and Debriefing the Activity

The feasible region for this problem is the shaded area in the following graph.

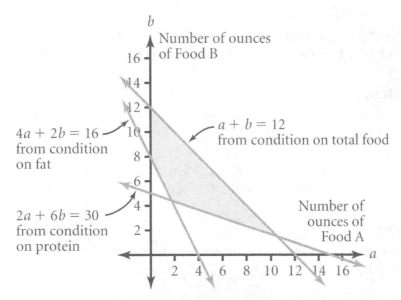

In drawing the feasible region, students need to take into account that the variables must be nonnegative, expressed using the inequalities $a \geq 0$ and $b \geq 0$. Without these constraints, the feasible region would continue across the vertical axis to the intersection, in the second quadrant, of the lines $4a + 2b = 16$ and $a + b = 12$.

If students do go outside the first quadrant, you can identify a point that fits the original three inequalities, such as $(-1, 11)$, and ask whether it makes sense for it to be in the feasible region.

Key Question

Should $(-1, 11)$ be in the feasible region? Is it a possible solution?

Picturing Pictures

Intent

This final activity in *Cookies and Inequalities* presents a new context in which students can bring to bear all of the tools they have developed in the unit.

Mathematics

In this activity, students will write inequalities to express the problem constraints; graph these constraint inequalities, including the implicit nonnegativity constraints, on a single set of axes to define the feasible region; write an equation for the profit (which has not been discussed since early in the unit); and find the profit for several points in the feasible region.

Progression

Students work on the activity individually and share their results in groups and with the class.

Approximate Time

5 minutes for introduction

25 minutes for activity (at home or in class)

20 minutes for discussion

Classroom Organization

Individuals, then groups, followed by whole-class discussion

Doing the Activity

Before beginning this activity, the class should agree on which variable to put on which axis. It is convenient, given the notes that follow, to place p on the horizontal axis and w on the vertical axis.

Discussing and Debriefing the Activity

Ask students how they represented the two constraints symbolically. They should have written inequalities equivalent to these.

$$p + w \leq 16 \qquad \text{(for the number of pictures)}$$

$$5p + 15w \leq 180 \qquad \text{(for the money available for materials)}$$

Students may have remembered to include the inequalities $p \geq 0$ and $w \geq 0$. If not, you can bring this up as you discuss the feasible region.

Ask for a volunteer to present the development of the feasible region, which is the shaded area in the graph here.

If students do not yet have a strong grasp of the connection between inequalities and their graphs, have students offer specific points that fit the constraints and then have the class check that they satisfy all of the constraints, including $p \geq 0$ and $w \geq 0$. Continue plotting points until the connection between the developing graph and the equations $p + w = 16$ and $5p + 15w = 180$ becomes clear.

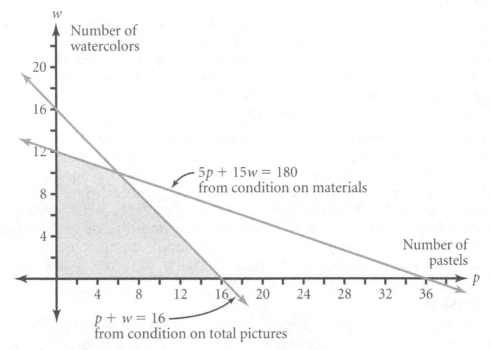

The Feasible Region for
Picturing Pictures

Ask, **What does the graph tell you?** Students should recognize that each point in the feasible region, or at least each whole-number point, represents a possible choice about how many pictures of each type Hassan can paint.

The issue of limiting the variables to whole numbers may arise again here. Students may point out that only the points with whole-number coordinates make sense in this context.

Have several students give the profit for one of the points they used in Question 3. You might put this information into an In-Out table as it is presented. Then ask for the profit equation in terms of the variables p and w, which should be equivalent to

$$Profit = 40p + 100w$$

You may want to look at the points students suggested and their associated profits to determine the maximum profit achieved so far. Tell students that they will soon learn a way to find the maximum profit and to prove that it is the maximum.

Key Question

What does the graph tell you?

Using the Feasible Region

Intent

The development of graphical methods for representing inequalities will be extended to representing a set of solutions to a linear programming problem.

Mathematics

The set of constraints in a linear programming problem with two variables consists of inequalities whose graphs are regions in the plane called **half planes**. The intersection of these half planes defines the **feasible region** for the system—the set of points whose coordinates satisfy all of the constraints simultaneously. The feasible region is a polygon in the plane, with the constraints defining its vertices and edges.

The goal of this problem is to identify which one of the possible combinations of plain and iced cookies also maximizes the bakers' profit, which is a function of the number of plain and iced cookies. In linear programming problems, the expression to be maximized or minimized is called the *objective function.* In *Using the Feasible Region,* students will learn that the solutions that produce a given profit are collinear and that by moving this **profit line** relative to the feasible region, keeping its slope constant, they can locate the vertex or edge of the feasible region that maximizes profit.

Progression

Using the Feasible Region presents four linear programming contexts. Students will graph the feasible region for each context and explore the graphical features of the optimization function. They will also develop a variety of methods for finding equivalent inequalities and intersection points for pairs of equations. In addition, students will present their results for the first POW of the unit and begin work on the second.

Profitable Pictures

Curtis and Hassan Make Choices

Finding Linear Graphs

POW 5: Kick It!

Hassan's a Hit!

You Are What You Eat

Changing What You Eat

Rock 'n' Rap

A Rock 'n' Rap Variation

Getting on Good Terms

Going Out for Lunch

Profitable Pictures

Intent

This activity continues students' work from *Picturing Pictures,* explicitly addressing the maximizing of profit.

Mathematics

In a linear programming problem with two variables, the set of constraints may be written as a system of linear inequalities. In this problem, that system is

$$5p + 15w \leq 180$$

$$p + w \leq 16$$

$$p \geq 0$$

$$w \geq 0$$

The graphical solution to this system defines a **feasible region**, a collection of points that form a polygon in the plane. The profit is also a linear function of p and w.

$$\text{Profit} = 40p + 100w$$

The central ideas of this activity are (1) that the set of points that produce a particular profit lie on the same line and (2) that the maximum profit is found by sliding this **profit line** up (increasing the profit) until it reaches the edge of the feasible region.

Progression

Students work on the activity and compare results in groups and as part of an extensive class discussion.

Approximate Time

55 minutes

Classroom Organization

Groups, followed by whole-class discussion

Materials

Profitable Pictures blackline master

Doing the Activity

Have students work in groups on the activity. Remind them that although they will work in groups, they will be preparing individual written reports.

If groups have trouble finding combinations that yield a specific profit, you might suggest that they consider points both outside and inside the feasible region.

For Question 5, you may have to ask questions to help groups formulate an explanation, such as, What do you notice about combinations that produce a given profit? What happens as the profit increases?

When most groups have had some time to explore Question 5, begin the discussion.

Discussing and Debriefing the Activity

Start the discussion by projecting the transparency of the graph from *Picturing Pictures* and having volunteers each mark a point from Question 2 to show a way in which Hassan can earn exactly $1,000. The only whole-number points in the feasible region that give this profit are (0, 10), (5, 8), and (10, 6).

Then have various students mark their points for Question 3, using a different color. The whole-number points in the feasible region that give a profit of exactly $500 are (0, 5), (5, 3), and (10, 1).

Finally, turn to Question 4. There are four whole-number points in the feasible region that yield a profit of $600: (0, 6), (5, 4), (10, 2), and (15, 0).

The Points for Each Profit Are Collinear

After all three sets of points have been plotted, ask the class, What do you notice about the different sets of points? Students should recognize that the points for each amount of profit lie on a straight line.

Introduce the term **profit line** for the set of points with a given profit. Have students draw in the complete lines on their graphs, connecting the individual points they found in Questions 2 through 4 and labeling each line with the profit it represents. On the transparency, you can extend the profit lines to include points outside the feasible region. Note that the profit lines include points whose coordinates are not whole numbers, even though Hassan can't make fractions of pictures.

Ask students, Why do the points for a given profit lie on a straight line? If necessary, ask what condition the coordinates must satisfy for a point to give a profit, for example, of $1,000.

Previously, students identified the profit expression as $40p + 100w$. Now they should realize that for a combination of pictures to yield a profit of $1,000, the number pair must satisfy the equation $40p + 100w = 1000$. In other words, the set of number pairs that give a profit of $1,000 is the same as the set of solutions to the equation $40p + 100w = 1000$, and the points corresponding to these pairs is the graph of the equation.

The diagram should now look something like the following.

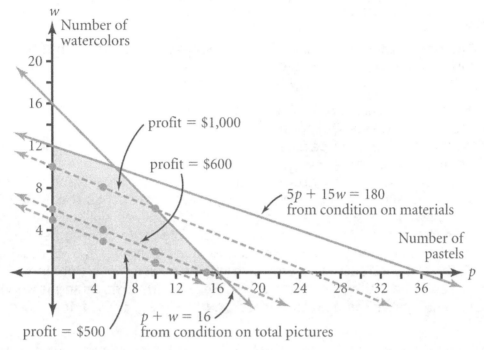

Profit Lines Are Parallel

A crucial element of this analysis is that the profit lines are all parallel. If it hasn't yet been mentioned, ask, **What do you notice about the set of profit lines?** Students should recognize that these are parallel lines and that as the profit increases the line "slides" upward to the right.

What does it mean for lines to be parallel? Bring out the key idea that parallel lines have no points in common.

Ask students how the algebraic representations of the three profit lines reinforce the visual evidence that the lines are parallel.

$$40p + 100w = 1000 \qquad \text{(for a profit of \$1,000)}$$
$$40p + 100w = 500 \qquad \text{(for a profit of \$500)}$$
$$40p + 100w = 600 \qquad \text{(for a profit of \$600)}$$

Students may reason about this in various ways. One approach is to observe that the equations $40p + 100w = 1000$ and $40p + 100w = 500$ cannot have any solutions in common; for any values of p and w, the expression $40p + 100w$ cannot equal both 1000 and 500. Because these linear equations have no common solutions, the graphs have no common points, which means their lines must be parallel.

Maximizing Profit

Now ask the class, **What is Hassan's maximum possible profit? How can you be sure?** Roughly, students' reasoning should begin with

these ideas:

- The points that give a specific profit satisfy an equation that has a form similar to those above.

$$40p + 100w = \text{Profit}$$

- For any particular profit, these points will lie on a line that is parallel to the profit lines for Questions 2 through 4.
- Within this family of parallel lines, profit increases as the line moves up and to the right.

From here, students should intuitively understand that they want to "slide" one of these parallel lines up and to the right until it reaches the edge of feasible region. If the sketch is made carefully, students will see that among those lines in the family that intersect the feasible region, the most "extreme" line is the one through the point where the lines $p + w = 16$ and $5p + 15w = 180$ intersect. Therefore, the point where these two lines intersect represents the maximum profit. Because this reasoning is so visual, most students should be able to understand it, even if they didn't discover it on their own.

At least some students are likely to have found the coordinates of this point, (6, 10), either by guess-and-check or some other means. Ask, **How can you confirm that you have the right coordinates for the intersection point?** Have students verify that this point fits both equations. Then ask them to compute the profit for this combination of pictures, perhaps comparing it to any values they previously thought were optimal.

Because the profit lines are almost parallel to the line $5p + 15w = 180$, it may not be clear from students' graphs where the family of parallel lines leaves the feasible region. If so, ask, **If it's not clear from the graph, how can you decide which point maximizes profit?**

You might ask what other points seem likely. Students should realize from the graph that the point (0, 12) is also a reasonable candidate. **How can you be sure which is the right point if you don't trust your graph?** Students should recognize that they could compute the profit at the two points and compare. The point (6, 10), representing 6 pastels and 10 watercolors, gives a profit of $1,240, while the point (0, 12), representing 12 watercolors, gives a profit of only $1,200.

When Profit Lines Are Parallel to a Constraint Line

When the family of parallel lines is parallel to a side of the feasible region, all points on that side will give the same profit. You can have an interesting discussion about how one would make a decision in that case.

For example, Hassan might choose a point along that side based on what he likes to paint, because the profit is the same. In the unit problem, if one side of the feasible region were parallel to the family of parallel lines, the Woos might choose the point along that side that maximizes the number of plain cookies, because they think plain cookies are healthier.

In discussing the family of parallel lines, you might remind students that **slope** is a mathematical term related to "the amount of slant" a line has. Because parallel lines have "the same amount of slant," they have the same slope.

Finding the Point of Intersection

Students have realized that the point of maximum profit is the point where the lines $p + w = 16$ and $5p + 15w = 180$ meet. Ask them to share their methods for finding the coordinates of this intersection point.

These equations are simple enough that guess-and-check probably worked for finding the values of 6 and 10 for p and w, respectively. Encourage students to use guess-and-check as a reasonable first approach, but point out that in other situations, the method might not suffice, especially if the solution involves fractions.

Another good approach is to estimate from the graph. Because the coordinates are whole numbers in Hassan's problem, this method would give the exact values.

Finally, students can solve for the intersection symbolically, using methods they began developing in *The Overland Trail* with the help of the "mystery bags" model.

If students find the numbers by guess-and-check, urge them to check that the answer is a good approximation of the coordinates of the point of intersection on the graph. If they find the numbers by estimating the coordinates from the graph, they should check that the numbers do indeed satisfy both equations.

Whatever method they use, bring out that the values for p and w have two distinct but closely related properties:

- These numbers are the p- and w-coordinates of the point where the two lines intersect.
- These numbers are the values for p and w that satisfy both equations.

Key Questions

What do you notice about combinations that produce a given profit?

What happens as the profit increases?

What do you notice about the different sets of points?

Why do the points for a given profit lie on a straight line?

What do you notice about the set of profit lines?

What does it mean for lines to be parallel?

What is Hassan's maximum possible profit? How can you be sure?

How can you confirm that you have the right coordinates for the intersection point?

If it's not clear from the graph, how can you decide which point maximizes profit?

Curtis and Hassan Make Choices

Intent

This activity reinforces students' understanding of linear functions.

Mathematics

In this activity, students will find sets of inputs that produce specific outputs for two functions:

$$\text{Cost} = 2A + 3B \qquad \text{(based on } \textit{Healthy Animals} \text{ and } \textit{Healthy Diets}\text{)}$$

$$\text{Profit} = 50p + 175w \qquad \text{(based on } \textit{Picturing Pictures} \text{ and } \textit{Profitable Pictures}\text{)}$$

The conclusion they should draw in each case is that the points that produce a particular cost (or profit) are collinear.

Progression

Students work on this activity individually, likely for homework during the classwork on *Profitable Pictures.* They review their work as part of the discussion of that larger activity.

Approximate Time

20 minutes for activity (at home or in class)

15 minutes for discussion

Classroom Organization

Individuals

Doing the Activity

This activity requires little or no introduction.

Discussing and Debriefing the Activity

Review this activity as part of the discussion of *Profitable Pictures* in connection with parallel profit lines.

Finding Linear Graphs

Intent

This activity focuses students' attention on methods for graphing linear equations.

Mathematics

Beginning in *The Overland Trail,* students have been devising methods for solving equations. Given the focus on linear equations in this unit—and, in particular, the need to find the intersection points that form the vertices of feasible regions in order to maximize profit—this activity gives students the opportunity to share the methods they have come to understand and use for solving linear equations.

Progression

Students work on the activity individually and share their results in groups and with the class.

Approximate Time

20 minutes for activity (at home or in class)

15 minutes for discussion

Classroom Organization

Individuals, then groups, followed by whole-class discussion

Doing the Activity

This activity requires little or no introduction.

Discussing and Debriefing the Activity

Focus the discussion on Question 3 by asking students to share in their groups their ideas for graphing linear equations. After this group discussion, ask each group to present at least one idea to the whole class.

Students should come to the general idea that they draw a graph by finding some points that satisfy the equation and then connecting those points with a straight line. Some may realize that two points will suffice. Others may be aware that it's often good to use a third point as a check.

If no one brings up the idea that the easiest points to find are often the two intercepts, ask, How do you come up with specific points to plot?

Students need not come away with any particular method for graphing a linear equation, as long as they know at least one method.

Key Question

How do you come up with specific points to plot?

POW 5: Kick It!

Intent

This POW will help emphasize the difference between conjectures and proved conclusions.

Mathematics

In this activity, football scores are determined by combining 5-point field goals and 3-point touchdowns. Students investigate the scores that are possible to obtain by combining multiples of these numbers and then look for generalizations involving other scoring systems. This problem is a special case of the general class of number theory problems known as "postage stamp" problems.

In the opening example, students will find that these scores are possible:

$$3, 5, 6, 8 \ (5 + 3), 9 \ (3 \bullet 3), 10 \ (2 \bullet 5), \ldots$$

The scores 1, 2, 4, and 7 are not possible. In addition, all scores from 8 on are possible, because the scores from 8 through 10 are possible, and all other numbers can be found by adding multiples of 3 to one of these three numbers.

Students will experiment with other scoring systems, looking for patterns. In some cases, there is a highest impossible score, whereas in others there is not. A general rule for finding the highest impossible score is $ab-(a + b)$, where a and b are the two score possibilities and a and b are relatively prime.

Progression

After an initial discussion, students work on the activity outside of class.

Approximate Time

10 minutes for introduction

1–3 hours for activity (at home)

20 minutes for presentations

Classroom Organization

Whole class, then individuals, followed by whole-class presentations

Doing the Activity

You may want to assure students that they do not need to know anything about football to work on this POW. They only need to know that in this activity, there are two ways to score points: field goals (worth 5 points) and touchdowns (3 points). Students will be looking for patterns among the possible scores and are expected to prove some of their results. Be sure they are aware, for example, that Question 1 involves proving that all scores above a certain value can be achieved using field goals and touchdowns. This is a good activity with which to make the distinction between showing lots of examples and giving a general proof.

Discussing and Debriefing the Activity

For Question 1, presenters presumably will have found that the highest impossible score is 7. Focus on their reasoning as to why every higher score is possible.

If they are having difficulty offering a convincing explanation, suggest that they find a sequence of consecutive scores that are possible and then note how every larger score can be obtained by combining scores in that sequence.

For example, they might show explicitly that all scores from 10 through 19 are possible and then observe that they can obtain any larger score by adding 10 points (2 field goals) as often as necessary to one of these scores. For instance, they can get a score of 47 by taking the scoring combination that gives 17 and adding 6 additional field goals. (More efficient variations of this proof are possible, but using multiples of 10 is comparatively easy for students to understand.)

As presenters discuss Questions 2 and 3, they will be offering conclusions either in connection with specific other scoring systems or as general principles. As a class, list these conclusions as they are presented, labeling each one as either "Proven Conclusions" or "Conjectures." For example, if a presenter gives a proof of a statement, list it under "Proven Conclusions." If a presenter offers a conclusion without satisfactory proof, list it under "Conjectures." Let students take the lead in deciding which statement goes in which list, intervening if they think that an incorrect statement has been proved or if you believe that the explanation is insufficient. When in doubt, it's probably better to err on the side of caution, labeling the statement "Conjecture."

As an extension, students might consider the situation of more than two kinds of scores—for example, field goals are worth 6 points, touchdowns 10 points, and safeties 15 points. (The numbers in this particular example have the peculiar property that any two of them have a common factor greater than 1, but no integer greater than 1 divides all three.)

Sample Conclusions and Proofs

Here are some conclusions and proofs that might arise. It may be best not to read these until you have worked on the problem yourself. These ideas are intended as background information for you, rather than as expectations of what students will do.

Example 1: This example concerns a case in which there is no highest impossible score.

Conclusion: If the number of points for a field goal is 2 and the number of points for a touchdown is 8, then there is no highest impossible score.

Proof: You can get only even scores in this game, so every odd number is impossible. Because there is no highest odd number, there is no highest impossible score.

The familiar terminology of even and odd makes this example easier to state than a case in which the scores are, for example, 5 and 15. If you get several conclusions of this type, you may want to move students toward this more general statement of the conclusion:

Conclusion: If the number of points for a field goal is more than 1 and divides the number of points for a touchdown, then there is no highest impossible score. (This conclusion is stated in terms of the score for a field goal dividing the score for a touchdown, but it would work just as well the other way around.)

Proof: All the scores made only with field goals are multiples of the points for one field goal. Because the points for each touchdown are also multiples of that number, then all possible combinations of the two types of score are multiples of that number. Therefore, every score that is not a multiple of that number is impossible. Because we can always find a number greater than any given number that isn't a multiple, there is no highest impossible score.

Example 2: This example concerns a family of cases in which there is a highest impossible score. The reasoning is similar to that for the case presented in the POW.

Conclusion: If the number of points for a field goal is 2 and the number of points for a touchdown is odd, then the highest impossible score is 2 less than the score for a touchdown.

Proof: You can get any even score from field goals and any odd score at or above the value of a touchdown by using one touchdown and the right number of field goals. But you can't get any odd numbers less than the value of a touchdown.

Key Questions

How do you prove your conclusion?

What is a conjecture?

Supplemental Activity

Kick It Harder! (extension) encourages students to develop further generalizations and proofs for this class of problems.

Hassan's a Hit!

Intent

Students return to the scenario in *Picturing Pictures* and *Profitable Pictures* to search for the maximum profit using a different profit function.

Mathematics

Students often believe that once they have graphed the feasible region, they can maximize profit by finding the point in the region that is farthest from the origin. Through this activity, they will learn that changing the profit function can change the point that gives the maximum profit, even when the constraints remain the same, and that they must draw the profit lines to determine the maximum profit.

Progression

Students work on this activity in groups and share discoveries in a class discussion.

Approximate Time

35 minutes

Classroom Organization

Groups, followed by whole-class discussion

Materials

Students' work from *Picturing Pictures* and *Profitable Pictures*

Doing the Activity

Have students work in groups on the activity. You might ask a few groups to prepare presentations for the class discussion.

Discussing and Debriefing the Activity

Ask a few groups to make presentations. The feasible region is the same as before, but now the profit function is

$$\text{Profit} = 50p + 175w$$

As in the original problem, students can compare profits at specific points to confirm their graphical analyses. In this case, the profit at (6, 10) is $2,050, while the profit at (0, 12) is $2,100. The graph shows the feasible region along with three of the new profit lines.

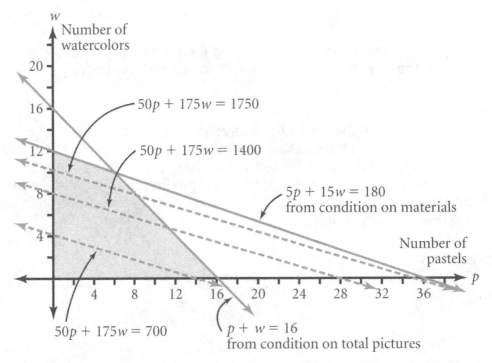

Profit Lines for
Hassan's A Hit!

$50p + 175w = 1750$

$50p + 175w = 1400$

$5p + 15w = 180$
from condition on materials

Number of
pastels

Number of
watercolors

$50p + 175w = 700$

$p + w = 16$
from condition on total pictures

By drawing these profit lines on the original feasible region, students will probably notice that they seem less steep than the graph of the equation $5p + 15w = 180$. Thus, as the profit increases, the last profit line to touch the feasible region does so at (0, 12), not at (6, 10) as in the original problem.

It is important that all students recognize that changing the slope of the parallel profit lines can change the maximum point in the feasible region. Thus, drawing the feasible region does not by itself give the answer. To drive this point home, ask, **Shouldn't the point giving the maximum profit be the point in the region farthest from the origin?**

You might have students experiment to note how changing the profit amounts for each type of picture affects the combination of pictures that gives the maximum total profit. They might also try to find profit values for which the maximum total profit is at (16, 0).

Key Question

Shouldn't the point giving the maximum profit be the point in the region farthest from the origin?

You Are What You Eat

Intent

A new linear programming situation allows students to further develop their abilities to identify constraints, write inequalities, graph feasible regions, and optimize a value based on a feasible region.

Mathematics

In this activity, students are asked to minimize the amount of cereal needed to meet two constraints (other than those involving nonnegativity): one to ensure enough protein, the other to limit carbohydrates. One of the constraints ends up having no influence on the answer.

Progression

Students work on the activity individually and share results in groups and with the class.

Approximate Time

20 minutes for activity (at home or in class)

20 minutes for discussion

Classroom Organization

Individuals, then groups, followed by whole-class discussion

Doing the Activity

If you think it is necessary, have volunteers read the activity aloud and identify the variables in the situation.

Discussing and Debriefing the Activity

Have students share their findings in groups. Ask each group to decide on the combination with the least amount of cereal that will satisfy Mr. Hernandez. Then choose one group to present its solution, which should answer the question, **What combination of cereals with the least total amount will satisfy Mr. Hernandez?**

Students will probably set this up like previous problems. If they use x for ounces of Fruit-Nuts and y for ounces of Crispies, the constraints are as follows.

$2x + y \geq 5$ (to guarantee enough protein)

$15x + 10y \leq 50$ (to avoid too much carbohydrate)

$x \geq 0, y \geq 0$ (because the amounts can't be negative)

Students should also note that the twins' goal is to minimize $x + y$.

The shaded area in the graph below shows the feasible region for this problem. The dashed line is a sample "consumption line" showing combinations for which the twins eat a total of 1 ounce of cereal.

The Feasible Region and a Consumption Line for *You Are What You Eat*

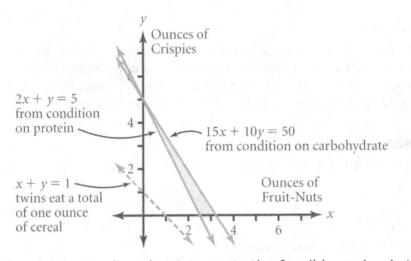

The "lowest" consumption line that intersects the feasible region is the one that goes through the point $(2\frac{1}{2}, 0)$, which is where the line $2x + y = 5$ meets the x-axis. Therefore, $2\frac{1}{2}$ ounces of Fruit-Nuts (and no Crispies) would be the best solution from the twins' point of view.

Point out that the carbohydrates constraint does not affect the solution and ask, **Why doesn't the carbohydrates constraint play a role in the solution?** Students should say something to the effect that the limit of 50 grams of carbohydrates is high enough that the twins get their protein requirement without approaching the carbohydrates limit. Because the twins want to eat as little as possible anyway, the carbohydrates condition is not an issue.

Students may come up with explanations for this solution without making a graph or formally identifying $x + y$ as the quantity to be minimized. For example, they may point out that the twins get more protein per ounce from Fruit-Nuts than Crispies, so there is no reason to eat the latter. If they present this reasoning, they need to check that $2\frac{1}{2}$ ounces of Fruit-Nuts does not provide too much carbohydrate.

Ask students if they recall encountering anything similar in the unit problem. **Did anything like this happen in the unit problem?** If necessary, refer them to the graph of the feasible region for the Woos. The line $P + I = 140$, from the oven-space constraint, is completely outside the feasible region.

Key Questions

What combination of cereals with the least total amount will satisfy Mr. Hernandez?

Why doesn't the carbohydrates constraint play a role in the solution?

Did anything like this happen in the unit problem?

Changing What You Eat

Intent

Students consider two variations on the problem posed in *You Are What You Eat.*

Mathematics

The two variations offered here change the constraints in two different ways, resulting in two new feasible regions. In each case, students are to minimize consumption. In one example, one of the constraint lines is parallel to the consumption function, so a set of points, rather than a single point, will minimize the consumption value.

Progression

Students work on the activity individually. In a class discussion, they review the collection of linear programming situations that they have explored so far and observe commonalities in their solution methods.

Approximate Time

20 minutes for activity (at home or in class)

40 minutes for discussion

Classroom Organization

Individuals, followed by group and whole-class discussion

Doing the Activity

This activity will require little or no introduction.

Discussing and Debriefing the Activity

You may want to let students spend a few minutes in their groups comparing results and then have representatives present their ideas. They should be able to summarize how to use a feasible region and a family of parallel lines to find a maximum or minimum.

After some specific combinations have been suggested for Question 1, you might ask, **Will *any* $2\frac{1}{2}$-ounce combination work?** Students should recognize that any $2\frac{1}{2}$-ounce combination will do, as any such combination is a minimum amount of cereal that provides enough protein and not too much carbohydrate.

The diagram below shows the feasible region for Question 1. Because one of the constraint lines, $2x + 2y = 5$, is parallel to the family of consumption lines, any point along this line represents the minimum amount of cereal.

The Feasible Region for
Changing What You Eat, Question 1

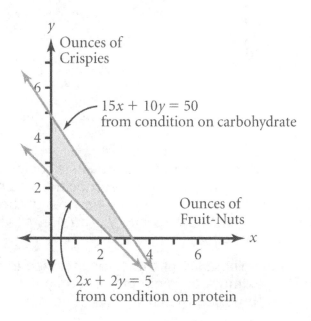

For Question 2, you might ask, **Why can't the twins simply eat $1\frac{2}{3}$ ounces of Fruit-Nuts?** Though it provides enough protein with the minimum amount of cereal, $1\frac{2}{3}$ ounces of Fruit-Nuts has too much carbohydrate. Therefore, the twins need to include some Crispies, which has less carbohydrate per ounce. It turns out that exactly 1 ounce of each cereal meets the requirements with the least total cereal, as the following diagram of the feasible region shows.

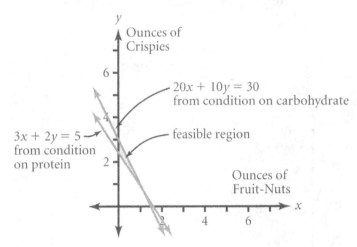

Now that students have worked with several feasible regions, it's a good time to ask them to summarize what they have been doing. You might first have them list the various situations for which they have used the same basic reasoning: *Profitable Pictures, Hassan's a Hit!, You Are What You Eat,* and the two situations in this activity.

Then, ask students to develop a list of the steps involved in using a family of parallel lines to find a maximum or minimum. You might have them work in groups for a while and then bring the class back together.

Here is a sample of what to look for, phrased in terms of profit lines. A list for consumption lines (such as for the breakfast problems) would be similar.

- Graph the feasible region.
- Graph some combinations for a given profit to get a straight line.
- Vary the profit to get a family of parallel lines.
- See that as profit increases, the parallel lines shift up and to the right.
- Look for a line in the family that is farthest "up and to the right" but that still intersects the feasible region.
- Identify the point where this line crosses the feasible region. This is the desired point.

Key Questions

Will any $2\frac{1}{2}$-ounce combination work?

Why can't the twins simply eat $1\frac{2}{3}$ ounces of Fruit-Nuts?

Supplemental Activity

More Cereal Variations (extension) poses two more questions about the breakfast situation.

Rock 'n' Rap

Intent

This activity presents students with a linear programming situation that has more than two constraints.

Mathematics

Students continue to develop linear programming tools for solving real-world situations. In this activity, they will write constraint inequalities, graph them to find the feasible region, explore a profit function, find the maximum profit, and explain their methods for doing so. They will review that the point of intersection of two lines is the same as the common solution to the two equations.

Progression

Students do this activity independently, applying the concepts and techniques they have developed so far, and share their work in groups and a class discussion. After students explore *A Rock 'n' Rap Variation*, the class will return to this activity and investigate how to solve the problem on the graphing calculator.

Approximate Time

10 minutes for introduction

20 minutes for activity (at home or in class)

20 minutes for discussion

30 minutes for exploring the problem on the graphing calculator

Classroom Organization

Individuals, then groups, followed by whole-class discussion

Doing the Activity

Allow about 10 minutes for students to read the activity and ask questions about any parts they find unclear. They may find the condition that the music company will "not release more rap music than rock" confusing, as it is stated negatively. You may want to suggest that they examine some numeric cases to help them determine how to express this condition as an inequality.

Ask the class, **What are the constraints in the problem?**

Agree as a class on what variables to use and which axes to put them on. The discussion that follows uses *x* to represent the number of rock CDs produced and *y* to represent the number of rap CDs produced.

Discussing and Debriefing the Activity

Have students share their findings in their groups. They should have come up with constraints equivalent to these.

$15x + 12y \leq 150$	(for the amount of money available for production)
$18x + 25y \geq 175$	(for the amount of production time)
$y \leq x$	(as promised to the distributor)
$x \geq 0, y \geq 0$	(because the numbers can't be negative)

Have volunteers share with the class the graph and the feasible region, showing how the inequalities lead to the desired region.

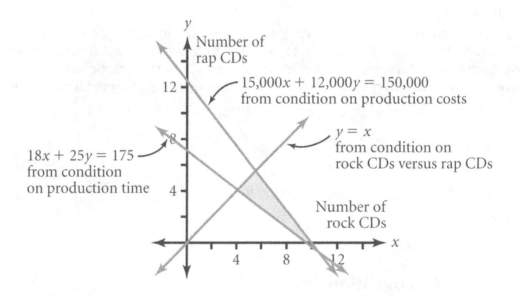

The Feasible Region for
Rock 'n' Rap

Now ask the class to identify the function that represents the profit in terms of x and y.

$$\text{Profit} = 20{,}000x + 30{,}000y$$

Have someone mark the points for Question 2a on the graph. The points (6, 0), (3, 2), and (0, 4) are the only points in the first quadrant (including its boundaries) with whole-number coordinates that give a profit of $120,000.

Ask another student to mark the points for Question 2b in a different color. The points (12, 0), (9, 2), (6, 4), (3, 6), and (0, 8) are the only points in the first quadrant with whole-number coordinates that give a profit of $240,000.

Then ask one or more students to explain how to find the point on the graph that will maximize profit. **What point maximizes profit?** This explanation involves two aspects:

- Explaining why the desired point is at the intersection of the graphs of $y = x$ and $15,000x + 12,000y = 150,000$
- Finding the coordinates of this point of intersection

Building on their work in Question 2, students will probably employ the "family of parallel lines" reasoning. The graph below shows the feasible region and the profit lines $20,000x + 30,000y = 120,000$ and $20,000x + 30,000y = 240,000$.

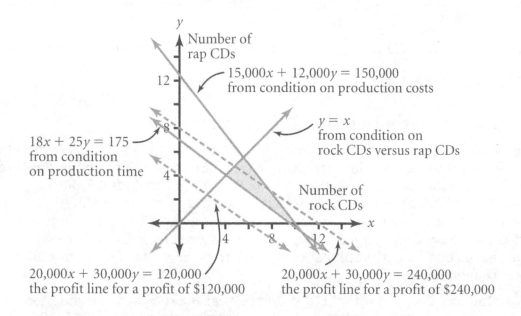

Profit Lines for the Feasible Region of
Rock 'n' Rap

How does the profit line change as the profit increases? Students should observe that as profit increases, the profit line moves up and to the right, and last touches the feasible region at the point where the lines $y = x$ and $15,000x + 12,000y = 150,000$ intersect.

How do you find the coordinates of the point of intersection? Some students will approach this graphically (by drawing careful graphs and estimating the coordinates or by using a graphing calculator), while others will reason symbolically. (Because one of the equations is $y = x$, the algebra is simple. Students will examine the general problem of finding coordinates of points of intersection in the activity *Get the Point.*)

The point of intersection has coordinates $(5\frac{5}{9}, 5\frac{5}{9})$, which students might approximate as (5.6, 5.6). The issue of using decimals to approximate fractional answers is addressed later. Get students to articulate the fact that $(5\frac{5}{9}, 5\frac{5}{9})$ represents both of the ideas given here.

- The coordinates of the point where the lines meet
- The common solution to the equations $y = x$ and $15{,}000x + 12{,}000y = 150{,}000$

Check that students are keeping in mind the connection between an equation and its graph. They should be aware that a point is on a graph if and only if its coordinates satisfy the equation. Thus, in this activity, the point where the lines meet has coordinates that satisfy both equations.

Is $5\frac{5}{9}$ a reasonable number of CDs? Get students to articulate that one can't buy $\frac{5}{9}$ of a CD, but that it is reasonable to spend the time and money to produce $\frac{5}{9}$ of a CD in a particular time period, and that this is explicitly permitted in the activity instructions.

Using a Calculator to Graph the Feasible Region

After students have finished *A Rock 'n' Rap Variation,* lead a class exploration of using a graphing calculator to solve this problem. Students have observed that the solution to this problem is the point where the lines $15{,}000x + 12{,}000y = 150{,}000$ and $y = x$ intersect. They may have found the coordinates either by plotting pencil-and-paper graphs or by using guess-and-check. Now they will learn how to use the calculator to find these coordinates and to get graphical confirmation of how the "family of parallel lines" reasoning works.

Have students begin by graphing the lines that make up the boundaries of the feasible region. They will first have to rewrite the equations $15{,}000x + 12{,}000y = 150{,}000$ and $18x + 25y = 175$ in "$y =$" form. Once they graph these rewritten equations and the equation $y = x$, they can see the feasible region. Adjustments in the viewing rectangle may be needed.

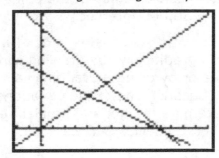

Next, have students look at the parallel lines for different profits. Again, they will have to rewrite the equations in "$y =$" form. For example, the profit equation $20{,}000x + 30{,}000y = 120{,}000$ becomes $y = \dfrac{\left(120{,}000 - 20{,}000x\right)}{30{,}000}$ or the equivalent. Students can easily check out various profit lines by substituting for 120,000 in the equation.

Let students play with this for a while, varying the profit and seeing how the profit line moves. This should reinforce the idea that the point where the profit is greatest is at the intersection of the two lines $y = x$ and $15{,}000x + 12{,}000y = 150{,}000$. They can use the trace feature, perhaps combined with adjustments in the viewing rectangle, to get the coordinates of the desired point. Using the trace feature, rather than an "intersect" or "solve" feature, gives students a more visual sense of what the coordinates mean. Once students understand this clearly, they will likely find shortcuts on their own.

You may want to ask students if it is possible to shade the feasible region on the calculator. They can consult the graphing calculator manual for details.

Key Questions

What are the constraints in the problem?

What point maximizes profit?

How does the profit line change as the profit increases?

How do you find the coordinates of this point of intersection?

Is $5\dfrac{5}{9}$ a reasonable number of CDs?

A Rock 'n' Rap Variation

Intent

In this follow-up to *Rock 'n' Rap,* the constraints remain the same but the profit function changes.

Mathematics

Students recognize that the change in the profit function will change the slope of its line. As this line is slid up to increase the profit, the last point to touch the feasible region will be different from the one that solved the original problem.

Progression

Students work in groups on the activity, building on their discoveries in *Rock 'n' Rap.*

Approximate Time

25 minutes

Classroom Organization

Groups

Doing the Activity

As students work, choose a group or two to prepare presentations.

Discussing and Debriefing the Activity

Have representatives from the groups you selected make presentations. The presenters should point out that the feasible region is the same as in *Rock 'n' Rap,* so all that needs to be done is to examine the new family of profit lines.

As the next diagram shows, if the profits are reversed, the profit line has a different slope, and the last point to be touched by a profit line is the point where the line $15,000x + 12,000y = 150,000$ meets the x-axis. This point has coordinates $(10, 0)$, so the company should make only rock CDs, and its profit will be $300,000.

The $240,000 Profit Line for
A Rock 'n' Rap Variation

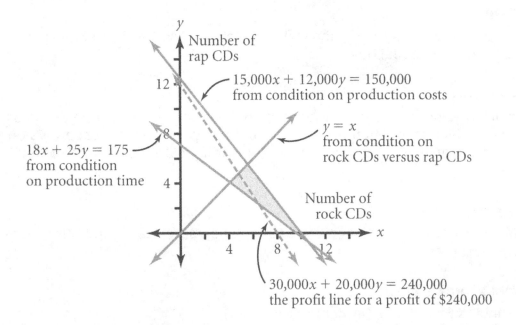

Key Question

How does changing the profit for each kind of CD affect the overall profit?

Supplemental Activities

Rap Is Hot! (reinforcement) presents another variation on the *Rock 'n' Rap* situation.

How Low Can You Get? (extension) provides a purely algebraic approach to the challenge of creating problem variations that yield different solutions.

© 2010 Interactive Mathematics Program

Getting on Good Terms

Intent

This activity reviews the symbolic manipulations used to find equivalent equations and inequalities.

Mathematics

This activity focuses on the process of solving for one variable in terms of another, which students must be able to do to graph certain equations on the calculator. Equivalent equations have the same solutions and the same graph.

Progression

Students work individually on the activity and share results in their groups and with the class.

Approximate Time

20 minutes for activity (at home or in class)

15 minutes for discussion

Classroom Organization

Individuals, then groups, followed by whole-class discussion

Doing the Activity

This activity requires little or no introduction.

Discussing and Debriefing the Activity

Have students compare answers in their groups. Then ask representatives from various groups to present their methods for solving for y in each equation.

Either partway through this process or after all the equations have been done, ask the class, **What relationship do the new "$y =$" equations have to the original equations?** As needed, remind students of the term **equivalent equations** and ask for a volunteer to explain what it means in terms of graphs. **How are the graphs of equivalent equations related?** Bring out that, by definition, equivalent equations have the same graph.

Ask students to describe the processes they used to find their equivalent equations. Students used the "mystery bags" model to begin to develop these processes in the Year 1 unit *The Overland Trail,* so you may want to ask them to use this model to explain their work.

Then ask students how they could use the idea that the equations should be equivalent to check their work. Help them realize that if they take any solution to a "y =" equation and substitute it into the original equation, it should satisfy the original equation.

Bring out that it is relatively easy to get solutions to a "y =" equation, because one can pick a value for x and substitute to get the corresponding value for y.

Have students try this for a couple of examples. For instance, they may have found the equation $y = \dfrac{\left(17 - 5x\right)}{3}$ for the equation $5x + 3y = 17$ in Question 3. You might have groups choose different values for x, substitute them into the expression $\dfrac{\left(17 - 5x\right)}{3}$ to get y, and then check that their number pairs fit the original equation.

In solving an equation for y in terms of x, some students may use decimal approximations rather than fractions. For example, in Question 3, they may solve $5x + 3y = 17$ and get $y = 5.67 - 1.67x$. If so, raise the point that this may lead to slightly inaccurate graphs or approximate solutions.

For instance, substituting 2 for x in this decimal approximation gives $y = 2.33$. Substituting 2.33 into the original equation, $5x + 3y = 17$, gives only an approximate solution; that is, $5 \cdot 2 + 3 \cdot 2.33 = 16.99$ instead of 17. Thus, $y = 5.67 - 1.67x$ is only approximately equivalent to $5x + 3y = 17$. Such inaccuracy may be unimportant in many contexts but important in others.

Key Questions

What relationship do the new "y =" equations have to the original equations?

How are the graphs of equivalent equations related?

Going Out for Lunch

Intent

In this, the final activity in *Using the Feasible Region,* students focus on writing and solving a system of equations, two important aspects of finding solutions to linear programming problems.

Mathematics

Students write two equations to express the conditions on two variables in a new context. Then they use informal methods, such as guess-and-check and graphing, to find the solution for this system.

Progression

Students work on this activity individually and discuss their solutions in groups and with the class.

Approximate Time

5 minutes for introduction

20 minutes for activity (at home or in class)

20 minutes for discussion

Classroom Organization

Individuals, then groups, followed by whole-class discussion

Doing the Activity

When you introduce the activity, tell students that they may have to use guess-and-check to find a solution. Assure them that intuition is a valid problem-solving strategy.

Discussing and Debriefing the Activity

Have students share their work in groups. Begin the discussion by having students suggest as many ways as possible to solve the problem without using equations.

Then ask, **Is this the only solution? Why do you think so?** One possible intuitive explanation is that the more hot dogs there are, the less it costs for 23 people, and the more hamburgers there are, the more it costs. This means there cannot be more than one solution. Whether students use this particular argument or another, try to get them to express their ideas as clearly as possible, and bring out that such an argument is, in fact, a proof of uniqueness.

After this intuitive discussion, ask a volunteer to show how the problem can be expressed symbolically, including how he or she defined the variables and wrote the equations.

Emphasize again that when we use variables to represent unknown numbers, we have to state exactly what the variables stand for. For instance, the statement "x = hot dogs" is unclear, as it may mean "the number of hot dogs" or "the cost of hot dogs."

If students use x for the number of hot dogs and y for the number of hamburgers, the equations might look like this:

$$x + y = 23$$

$$1.10x + 1.50y = 32.10$$

Once the equations have been presented, have students check that the solution arrived at informally satisfies them. Make sure they recognize that verification involves simply substituting 6 for x and 17 for y into the equations and checking that this produces true statements. (You do not need to discuss how to solve this pair of equations symbolically. Students will begin learning about that in the next activity, *Get the Point.*)

Tell students that when they have a set of equations using the same variables and are looking for values for those variables that satisfy all the equations, it is called a **system of equations** (or **system of simultaneous equations**). A set of values that fits all the equations is called a *solution* (or *common solution*) to the system.

In particular, the pair of equations students developed for this activity is called a *system of linear equations.* In the next activity, *Get the Point,* students will learn how to use algebraic methods to find the common solution to such a pair of linear equations.

Some students may have used a calculator to solve this activity. You might ask everyone to graph the two equations on their calculators to check whether they get the same solution.

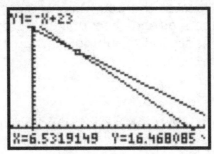

Key Questions

How could you solve this problem without using equations?

Is this the only solution? Why do you think so?

How would you express this problem algebraically?

Points of Intersection

Intent

The activities in *Points of Intersection* focus on an important skill for solving linear programming problems: solving a system of two equations in two unknowns.

Mathematics

In this unit, students add symbolic methods to their graphical and guess-and-check approaches to finding the solution to a system of linear equations. The activities and their discussions emphasize two such methods: (1) solving both equations for the same variable and setting the resulting expressions in one variable equal to each other and (2) solving one equation for one of the variables and substituting the resulting expression into the other equation.

Three possible cases arise when solving linear systems. If the equations are represented by lines in the plane that intersect in a single point, there will be one unique solution. Such a **system of equations** is said to be **independent**. If the lines are parallel, there are no solutions. This is an **inconsistent** system. Third, the two equations might have the same graph (that is, be **equivalent equations**), in which case any solution to one equation will also solve the other. Such **dependent** systems have an infinite number of solutions.

For many people, the preferred method of solving systems of equations in several unknowns is the elimination method, in which the equations are multiplied by constants and then added to or subtracted from one another to eliminate one variable at a time. Also known as Gaussian elimination, this method can be used to solve linear systems in two or more variables. The IMP curriculum does not present the elimination method yet, because the approach of solving both equations for y and setting the expressions equal is easier to conceptualize and emphasizes that each expression in x results in the same y. The process locates the point of intersection of the equations' graphs. In addition, students already recognize the idea of solving for y in terms of x as an important step in graphing on the calculator, and this approach builds on that technique.

Progression

In the first activity of *Points of Intersection,* students explore, share, and articulate symbolic methods for solving systems of equations. After several activities that solidify their understanding of these methods, the final activity raises the cases of inconsistent and dependent systems. In addition, students will complete their work on the second POW and begin work on the final POW of the unit.

Get the Point

Only One Variable

Set It Up

A Reflection on Money

POW 6: Shuttling Around

More Linear Systems

Get the Point

Intent

In this activity, students develop algebraic methods for solving pairs of linear equations. The goal is not for them to devise any particular method, but to find an algebraic procedure that works and makes sense to them. They may even find that they like one method for some systems and another method for others.

Mathematics

The solution to a system of two linear equations in two unknowns, if a unique solution exists, is the set of values for the variables that solve both equations simultaneously. It is the point in the plane at which the two equations' graphs intersect. Up to now, students have used a graphical approach or guess-and-check to find such solutions. In this activity, the focus is on developing symbolic methods.

The unit emphasizes two related symbolic approaches. Substitution involves solving one equation for one variable in terms of the other, substituting this expression into the second equation, and solving that equation for the remaining variable. The second approach is to solve each equation for the same variable and set the resulting expressions equal to each other. Both methods emphasize the meaning of the solution described above.

Progression

Students work in groups to solve five pairs of equations, check their results, and then describe the methods they used in general terms. Their work is followed by a class discussion of their methods. In addition, each group presents its work to the class.

Approximate Time

35 minutes for activity

30 minutes for discussion and presentations

Classroom Organization

Groups, followed by whole-class discussion and group presentations

Doing the Activity

In this activity, groups develop systematic algebraic ways to find the exact solution to a system of two linear equations. You may want to emphasize to students that an algebraic approach will always result in the exact solution, giving it an advantage over graphing. Also make sure students understand that they will produce individual written reports and that groups will make oral presentations.

Circulate as the groups work, asking questions as necessary. Questions 1a and 1b represent the simplest case as they involve equations with y already stated in terms of x. For Question 1a, you might ask, **What is true of all points on the line $y = 3x$?** or **How would you get the y-coordinate from the x-coordinate for a point on this line?** Try to elicit an answer to the effect that the y-coordinate is 3 times the x-coordinate for all points on the line.

Ask similar questions for the line $y = 2x + 5$ and then relate the two equations by asking what happens at the point of intersection. Bring out that students can find the x-coordinate by determining what value of x makes the two expressions equal, that is, by solving the equation $3x = 2x + 5$.

A similar approach will work for Question 1b.

Some groups may be clear on Questions 1a and 1b but not know what to do with the more complex problems. Point out that Questions 1a and 1b highlight the fact that in the simultaneous solution to the system, the y-values (as well as the x-values) must be the same for both equations. Then you might ask, **How could you use this same idea in Question 1c, and maybe in Question 1d?** Some students may revert to the previous method by solving both equations for y and setting them equal, but others may begin to experiment with substituting $4x + 1$, the equivalent of y, in the first equation of Question 1c. Encourage both methods as they emerge. This might also be a good time to emphasize the meaning of the term **equivalent equations**.

Question 2 asks students to articulate a general outline of their solution methods. Groups may come up with quite different methods, and you can encourage variety. Some groups may choose to describe more than one method and may suggest using one method in some cases and a different method in others.

Discussing and Debriefing the Activity

Have groups make their presentations on using algebra to find the coordinates of the point of intersection of two linear equations.

One possible method—which could be referred to as the "setting ys equal" method—might look like this.

1. Use each equation to get an expression for y in terms of x.
2. Set these two expressions equal to each other.
3. Solve this new equation, which has only one variable, x.
4. Substitute x into one of the original equations and solve for y.

Here is another possible outline for a method usually called substitution.

1. Solve one equation to get an expression for y in terms of x.
2. Replace y in the other equation with this expression.
3. Solve this new equation, which has only one variable, x.
4. Substitute x into one of the original equations and solve for y.

Key Questions

What is true of all points on the line $y = 3x$? How would you get the y-coordinate from the x-coordinate for a point on this line?

How could you use this same idea in Question 1c, and maybe in Question 1d?

Only One Variable

Intent

This activity connects students' work on *Get the Point* with methods for solving equations in one variable, which were developed beginning in the Year 1 unit *The Overland Trail.*

Mathematics

This activity reviews concepts for finding equivalent equations. It connects to students' work, beginning back in *The Overland Trail,* to develop ways to produce a new equation equivalent to a given equation. These techniques involved the distributive law and methods such as adding (or subtracting) the same term to both sides of an equation or multiplying (or dividing) both sides of an equation by the same nonzero term.

As students continue to consider linear programming situations, their facility with manipulating algebraic expressions is essential. In addition, this ability must be second nature for students to efficiently use the graphing calculator.

Progression

Students work individually on this activity and discuss their findings in class.

Approximate Time

20 minutes for activity (at home or in class)

25 minutes for discussion

Classroom Organization

Individuals, followed by whole-class discussion

Doing the Activity

Tell students that this activity will help them review ideas about solving linear equations in one variable.

Discussing and Debriefing the Activity

You might have students discuss Question 1 briefly in their groups before they present their work. Follow up each part by asking for alternative approaches.

Question 1a should be straightforward. You can use the presentation to review the basic principles for finding equivalent equations, perhaps referring to the mystery bags game from *The Overland Trail.*

If needed, use Question 1b to review the distributive property and methods for removing parentheses. Having students check the solution by substitution will also give you an opportunity to review the arithmetic of integers and the order of operations.

Questions 1c and 1d are designed to anticipate problems students may have in their continued work on *Get the Point*. You can take this opportunity to discuss how to work with equations involving fractions.

For Question 1c, some students might rewrite the equation as $0.5x + 1.5 = 29-2x$, which allows them to avoid the issue of fractions. Others might multiply both sides by 2 to get $x + 3 = 2(29-2x)$.

Question 1d is more complicated because fractions (with different denominators) appear on both sides of the equation. One approach is to eliminate the fractions by multiplying both sides by 4 and then by 6. Students may also realize that they can multiply directly by 24 or 12.

Have two or three volunteers share the problems they created for Question 2. Let the class try to write equations to represent the problems.

Set It Up

Intent

Students write a pair of linear equations to represent a situation and solve the system using methods developed in previous activities.

Mathematics

This activity combines several aspects of students' work so far in this unit: writing equations to represent real-world situations, solving equations using a graph, and solving equations symbolically. The activity also fosters algebraic thinking by asking students to make up a pair of equations that has a given solution.

Progression

Students work on the activity individually as they complete their work on *Get the Point*.

Approximate Time

5 minutes for introduction

20 minutes for activity (at home or in class)

25 minutes for discussion

Classroom Organization

Individuals, then groups, followed by whole-class discussion

Doing the Activity

When you introduce the activity, briefly review the relationship between symbolic and graphical methods for solving equations. Remind students to define their variables carefully.

Discussing and Debriefing the Activity

Have students share their work in groups. You may want to give transparencies to groups to prepare presentations on different parts of the activity: defining variables and writing equations, solving the equations symbolically, and solving the equations graphically.

You may want to have two or three presenters for Question 2, as many different linear systems have the given solution.

Question 1 should generate a pair of equations something like those below, where x represents the number of two-point shots and y represents the number of three-point shots. Focus on careful definition of variables—for example, x represents "the number of two-point shots," not "two-point shots."

$$x + y = 119$$
$$2x + 3y = 273$$

For the graphical approach, ask the presenter how he or she sketched the graphs. Use this opportunity to review ideas about graphing linear equations.

Similarly, for the symbolic approach, ask for details about what the presenter did. This presentation may give you some idea of what to expect in the presentations of *Get the Point.*

After presentation of graphical and algebraic methods, ask a volunteer to answer the original question of how many of each type of shot Marilyn made.

The Decimal Approximation Issue

The issue of decimal approximations, raised in *Getting on Good Terms,* may surface again here, because an approximate expression for y in terms of x may lead to a solution to the system of equations that is only approximate.

For instance, suppose students rewrite $x + y = 119$ as $y = 119 - x$ and rewrite $2x + 3y = 273$ as $3y = 273 - 2x$ and then as $y = 91 - 0.67x$. Setting the two expressions for y equal gives the equation

$$119 - x = 91 - 0.67x$$

Combining terms to get $28 = 0.33x$ and dividing both sides by 0.33 gives $x \approx 84.85$. Substituting this into the equation $x + y = 119$ gives $y \approx 34.15$. This is only an approximation of the exact answer, which is $x = 84$, $y = 35$.

If this issue arises, advise students to test their answers in both original equations. Substituting 84.85 for x and 34.15 for y into the expression $2x + 3y$ gives 272.15, not 273 as required.

For Question 2, ask presenters, How did you find a system of equations with this solution? (Note: The pair of equations $x = 3$, $y = 5$ is itself, technically, an answer to the question.)

If no one describes the method of starting with a linear expression in x and y and substituting to get the constant term, you might ask, Is there an equation you could use that has 7x + 9y on the left side?

As students present systems of equations with the given solution, raise the question of why there is more than one possible answer. How can you use graphs to explain why there are different systems with this solution? They should recognize that any pair of equations whose graphs are distinct lines through the point with coordinates $x = 3$, $y = 5$ will work. Bring out that there are infinitely many lines through this point, any pair of which will work.

Key Questions

How did you find a system of equations with this solution?

Is there an equation you could use that has 7x + 9y on the left side?

How can you use graphs to explain why there are different systems with this solution?

A Reflection on Money

Intent

This activity presents one more real-world situation that can be represented by and be solved using a system of linear equations. In addition, students solve several other systems, reinforcing their understanding of the methods they have been developing.

Mathematics

This activity continues work with real-world problems yielding systems of linear equations and methods for solving these systems. It asks students to use both graphical and symbolic approaches and then to reflect on the advantages and disadvantages of each.

Progression

Students work on the activity individually and share their work in groups and with the class.

Approximate Time

20 minutes for activity (at home or in class)

20 minutes for discussion

Classroom Organization

Individuals, then groups, followed by whole-class discussion

Doing the Activity

Remind students to clearly identify their variables as they translate Question 1 into words.

Discussing and Debriefing the Activity

Have students discuss the activity in groups. Solving these questions may be fairly routine by now, although the systems in Question 2 do have negative and fractional answers. Use your judgment about whether to have groups prepare presentations for either question.

Spend some time on Question 1c, perhaps asking several students to share advantages and disadvantages of the graphical and symbolic approaches.

POW 6: Shuttling Around

Intent

This third POW of the unit presents a puzzle that leads to a pattern that can be generalized.

Mathematics

This versatile problem can be solved in many ways: logically, visually, and algebraically. The various solutions allow students to become aware of connections among the methods, such as between a visual approach and an algebraic approach.

Progression

Students first explore the puzzle in class to understand its rules. Then, they work on the activity outside of class and present their findings in a week or so.

Approximate Time

10 minutes for introduction

1–3 hours for activity (at home)

20 minutes for presentations

Classroom Organization

Whole class, then individuals, followed by whole-class presentations

Materials

Game markers, such as counters of two colors or coins of two types

Doing the Activity

Perhaps the best way to introduce this POW is to let students work out the simplest case of the puzzle, which has one of each type of marker. They should quickly realize that this puzzle can be solved in three moves.

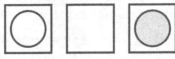

Remind students that they are to look for a general rule for the minimum number of moves needed.

Discussing and Debriefing the Activity

If students have developed a formula for the number of moves required, ask if anyone has a way to explain or prove it.

If no one presents a general formula, ask students how they might develop one. If needed, suggest compiling all the data students have collected into an In-Out table and looking for a rule. This may lead some students to recognize a formula. If not, you can leave it as an open question.

Number of Markers on Each Side	Moves Needed
1	3
2	8
3	.
4	.
5	.

Some students may solve this problem geometrically, keeping track of "slides" and "jumps." Separating "slides" and "jumps" in the Out of an In-Out table may help students discover the general formula.

Key Questions

How might you get a general formula for the number of moves?

How might you prove the general formula?

Supplemental Activity

Shuttling Variations (extension) is an excellent follow-up activity to this POW.

More Linear Systems

Intent

This activity introduces students to the range of possible outcomes when solving a system of linear equations.

Mathematics

Up to now, students have found a single point that solves each system of equations they have encountered. Such a system, called an **independent system**, consists of two lines that intersect at a single point.

This activity includes two examples illustrating other results students could obtain when solving a linear system. The two equations might produce lines that are parallel. Such an **inconsistent system** has no solutions. Or, the two equations might produce the same line. A **dependent system** such as this has infinitely many solutions: any points that solve one equation, and only those points, will solve the other.

Independent system

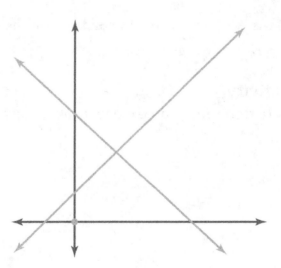

Inconsistent system

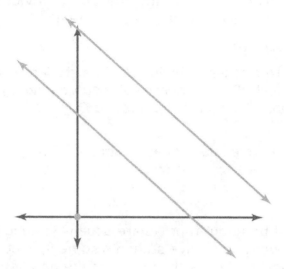

Progression

Students solve the questions in this activity individually. The two systems of equations that produce "weird" results will be the focus of the class discussion, which will bring out that a pair of linear equations may have no common solution and introduce the terms **inconsistent equations, dependent equations,** and **independent equations.**

Approximate Time

25 minutes for activity (at home or in class)

20 minutes for discussion

Classroom Organization

Individuals, then groups, followed by whole-class discussion

Doing the Activity

Let students discover the complications in these systems on their own and grapple with how to handle them. The follow-up discussion will explore inconsistent and dependent systems of linear equations.

Discussing and Debriefing the Activity

Let students compare answers in their groups. Then begin the discussion, perhaps by asking whether they had any difficulties with or any issues to raise about Question 1.

If no one points to Question 1d, ask about it yourself. Because simplifying each side leads to the equation $3w + 6 = 3w + 14$, the equation has no solution. You might remind students that they saw a similar situation in the activity *The Mystery Bags Game* in the Year 1 unit *The Overland Trail*.

You might take a similar approach with Question 2, reviewing Questions 2a and 2b only if students raise questions about them.

Inconsistent Systems

Ask for a volunteer to discuss Question 2c. The student will likely say something to the effect of his or her usual method leading to "something weird." You can probably connect that idea to Question 1d, which has no solution.

If students are unable to explain what's happening in Question 2c, ask, *What is happening in terms of the graphs of these equations? What usually happens when finding the common solution to a pair of linear equations?*

The response should bring out that we are usually looking for the point where two lines intersect, which may give students some insight. Have them graph the equations in Question 2c. As the two lines are parallel and thus have no points in common, the equations have no solution in common. Tell students that a pair of linear equations with no common solution is called an **inconsistent system**.

Dependent Systems

Point out that two lines will usually intersect in a unique point. Ask if there is anything else that can happen with two lines except for intersecting at a point or being parallel. As an example, give students the next pair of equations (which are the same as those in Question 2c except that the second constant term has been changed).

$$2x + 3y = 1 \text{ and } 6y = 2-4x$$

Students should recognize (by solving for y and graphing, if needed) that these equations have the same graph; they are **equivalent equations**. Point out that this might not have been obvious at a glance, because the equations have somewhat different forms. Bring out that the first equation is equivalent to $4x + 6y = 2$ (by multiplying it by 2), which is essentially the second equation.

Ask, *Are there any number pairs that fit both equations?* Students should realize that because the equations are equivalent, any number pair that fits one will fit the other, meaning there are infinitely many solutions to the system.

Explain that a pair of linear equations that have the same graph is called a **dependent system**, and that when two linear equations give distinct but intersecting lines, the equations are called **independent**.

It will probably be helpful to summarize the three possible outcomes for a pair of linear equations and match the algebraic result with the geometry. You might post these three statements.

- Unique solution (independent system) ⇔ intersecting lines
- No solution (inconsistent system) ⇔ parallel lines
- Infinitely many solutions (dependent system) ⇔ the same line

Key Questions

What is happening in terms of the graphs of these equations?

What usually happens when finding the common solution to a pair of linear equations?

What about the pair of equations $2x + 3y = 1$ and $6y = 2-4x$? Are there any number pairs that fit both equations?

Cookies and the University

Intent

In *Cookies and the University,* students solve the unit problem. They also reflect on the types of problems they have been solving in the unit.

Mathematics

The unit problem is a linear programming problem with six constraint inequalities. Students seek to find the point in the feasible region that maximizes a linear profit function. Because the optimum value is one of the feasible region's vertices, one of the key skills students will employ is solving a system of two linear equations with two unknowns. In *Cookies and the University,* students solve several additional systems of equations, as well as reflect on the distinction between such problems and linear programming problems.

Progression

Students begin *Cookies and the University* by solving the unit problem. They then work on several activities that help them distinguish between the two types of problems they have encountered in the unit. Finally, they create their own two-equation/two-unknown problems in anticipation of *Creating Problems,* in which they will create their own linear programming problems.

How Many of Each Kind? Revisited

A Charity Rock

Back on the Trail

Big State U

Inventing Problems

How Many of Each Kind? Revisited

Intent

This activity restates the unit problem, first presented in *How Many of Each Kind?* Students apply what they have learned over the course of the unit to solve the unit problem and then write a written report of their work.

Mathematics

In solving the unit problem, students bring to bear all of the skills and techniques they have acquired during the unit. They use constraint inequalities to define the feasible region for the situation. They find the coordinates of the vertices of the feasible region. They plot the profit function and then slide this line up to identify the optimal solution.

Progression

Students work in groups and individually to solve the problem. Each student then prepares a written report. The class discusses their findings.

Approximate Time

45 minutes for activity

20 minutes for discussion

Classroom Organization

Groups and individuals, followed by whole-class discussion

Doing the Activity

Tell students that they now have the tools they need to return to the central unit problem, first explored in *How Many of Each Kind?* Have them carefully review the details in the section "Your Task." Emphasize that their written reports should present their reasoning clearly and thoroughly.

Working in their groups, students will likely need at least 45 minutes to produce these reports. From their progress, decide how soon to begin presentations.

You may want to remind students that they did considerable work on this problem in the activities in *Cookies and Inequalities* and *Picturing Cookies.* They can use that earlier work as an aid, but their reports should explain everything from scratch. You may want to mention that their reports will be included in their portfolios.

Discussing and Debriefing the Activity

Have one or two groups present their analysis of the unit problem.

An important focus of the discussion should be on the family of parallel profit lines. The graph below shows the feasible region given earlier, with one of the profit lines—represented by a dashed line—added. This line, the graph of the equation $1.5P + 2I = 150$, shows those combinations of dozens of plain and iced cookies that give a total profit of $150. On their graphs, students should also label the coordinates of the key points shown—(30, 80) and (75, 50).

The Feasible Region for the Cookie Problem
with a $150 Profit Line

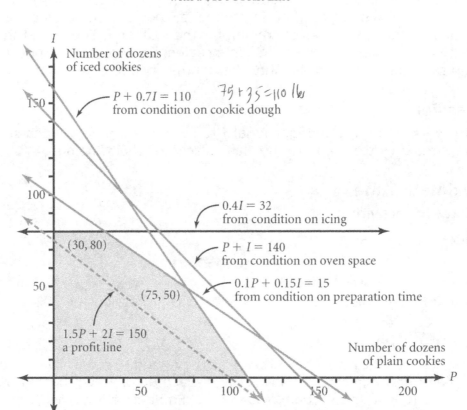

Ask some questions about the graph.

What does each line represent? How did you find the feasible region? Students should be able to explain what each of the lines represents and how the feasible region was determined. For example, they should be able to explain why the region includes the area below rather than above a certain line.

What expression describes profit? Students should be able to identify the expression $1.5P + 2I$ as describing the profit.

How do you know that (75, 50) is the best choice? Students will probably use the "family of parallel lines" reasoning to explain why the best choice for the Woos is to make 75 dozen plain cookies and 50 dozen iced cookies. The points that give any particular profit lie on a straight line. As profit increases, the line is replaced by a parallel one above and to its right. The last of these lines to intersect the feasible region is the one through the point (75, 50).

How can you be sure where the "parallel family of lines" leaves the region? For example, how do you know it isn't at the point (30, 80)? Students might respond that the answer must be at either (30, 80) or (75, 50), because these are corners, and then simply evaluate the profit expression $1.5P + 2I$ at both points. The profit at (30, 80) is $205.00, while the profit at (75, 50) is $212.50.

Key Questions

What does each line on the graph represent?

How do you find the feasible region?

What expression describes profit?

How do you know that (75, 50) is the best choice?

How can you be sure where the "parallel family of lines" leaves the region? For example, how do you know it isn't at the point (30, 80)?

A Charity Rock

Intent

In this activity, students review and practice writing and solving systems of equations.

Mathematics

Students reinforce their abilities to solve a system of equations, including recognizing **inconsistent** and **dependent systems**, and to define a system from the conditions in a problem context, including carefully defining variables.

Progression

Students work on this activity individually and share their results in groups and with the class.

Approximate Time

25 minutes for activity (at home or in class)

25 minutes for discussion

Classroom Organization

Individuals, then groups, followed by whole-class discussion

Doing the Activity

This activity requires little or no introduction.

Discussing and Debriefing the Activity

Part I of the activity contains three systems of equations to solve. In Part II, students develop a system from a problem context and then solve the system to answer the question. Use the discussion of Part I to review the concepts of inconsistent and dependent equations. Use Part II to focus on defining variables carefully and developing equations to describe a situation.

Part I: Solving Systems

Ask students to come to a consensus in their groups about the solutions to Part I. As groups are working, circulate to get a feel for how much time you need to spend reviewing this activity. It's probably worth at least having presentations on both Question 2 (an inconsistent system) and Question 3 (a dependent system).

Part II: Rocking Pebbles

Have one or two students present solutions to Part II. Make sure they define their variables clearly—for example, "x = the price of reserved-seat tickets" rather than "x = reserved seats."

Using x to represent the price of reserved-seat tickets and y to represent the price of general-admission tickets, students will derive a system of equations like the example here.

$$230x + 835y = 23{,}600 \qquad \text{(for the first night)}$$

$$250x + 980y = 27{,}100 \qquad \text{(for the second night)}$$

To solve the system, students could use the process they developed in *Get the Point*, graphing, or guess-and-check. Encourage discussion of the various methods and their connections.

Once students find the two ticket prices, they then need to determine the amount of the donation by calculating half the proceeds from the general-admission tickets.

Back on the Trail

Intent

This activity connects the work students have been doing in this unit to earlier work in the curriculum from the Year 1 unit *The Overland Trail.*

Mathematics

This activity contains two-equation/two-unknown problems drawn from the unit *The Overland Trail,* in which students first studied linear functions and found solutions to linear systems in informal ways. Now that students have formalized graphical and symbolic methods for solving linear systems, they can bring those skills to bear on these familiar tasks. The first problem will likely seem easier to them because one of the equations expresses one variable directly in terms of the other.

Progression

Students work individually on the two situations in this activity. The class discussion includes a look back at the two types of problems that students have been tackling in this unit.

Approximate Time

5 minutes for introduction

20 minutes for activity (at home or in class)

15 minutes for discussion

Classroom Organization

Individuals, followed by whole-class discussion

Doing the Activity

You might mention to students that now, using new algebraic techniques and methods they have developed since they first worked on the Year 1 unit *The Overland Trail,* they will re-solve problems they first solved in informal ways.

Discussing and Debriefing the Activity

Have one or two students present solutions to each problem. If neither presenter took a standard two-equation/two-unknown approach, ask for a volunteer to do the problem that way, making sure the variables are clearly defined.

In Part I, using b for the number of hours in a boy's shift and g for the number of hours in a girl's shift, the equations will likely be

$$b = g + 1$$

$$3b + 2g = 10$$

The solution to this system is $b = 2.4$, $g = 1.4$, so each boy's shift should be 2.4 hours and each girl's shift should be 1.4 hours.

For Part II, using a and c for the number of gallons of water used by an adult and a child, respectively, the likely equations are

$$3a + 5c = 15$$

$$2a + 4c = 11$$

The solution is $a = 2.5$, $c = 1.5$, meaning that each adult uses 2.5 gallons of water and each child uses 1.5 gallons.

Big State U

Intent

Through their work in this activity, students synthesize their ideas about how to solve complex problems with several constraints, preparing them to write one of their own.

Mathematics

This activity presents students with one final two-variable linear programming situation to set up and solve.

Progression

Students work on this final linear programming problem in groups and present their results to the class.

Approximate Time

25 minutes for activity

15 minutes for presentations

Classroom Organization

Groups, followed by whole-class presentations

Doing the Activity

When you introduce the activity, remind students to define their variables carefully and to be as clear as possible about why they are doing each step in the process.

Discussing and Debriefing the Activity

As groups conclude their work on this activity, choose one or two to make presentations.

Letting x represent the number of in-state students to be admitted and y represent the number of out-of-state students, the constraints are as follows:

$8000x + 2000y \geq 2{,}500{,}000$ (for the contributions the president wants)

$y \geq x$ (to satisfy the faculty's concern about grades)

$100x + 200y \leq 85{,}000$ (to fit the housing office's budget)

$x \geq 0, y \geq 0$ (because the values can't be negative)

The function that needs to be minimized, for the sake of the treasurer, is

$$\text{Educational cost} = 7200x + 6000y$$

The graphs of the equations that go with the constraints, as well as the feasible region (the small shaded triangle), are shown here.

The Feasible Region for *Big State U*

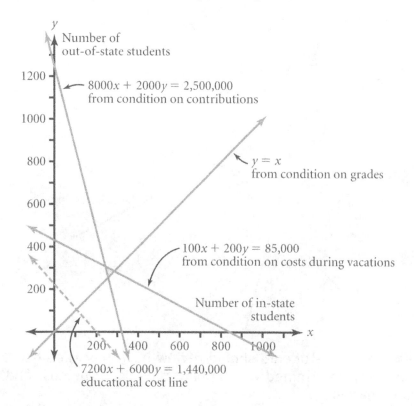

The dashed line is from the family of parallel lines that represent different educational cost conditions. This particular line, the graph of $7200x + 6000y = 1,440,000$, shows the possible enrollment combinations that give an educational cost of $1,440,000. The graph shows that the minimum "cost line" to intersect the feasible region is the one through the intersection of the lines $y = x$ and $8000x + 2000y = 2,500,000$.

The following graph shows the feasible region in more detail and gives the coordinates of the points of intersection. The point that minimizes educational cost is (250, 250). Thus, the school should admit 250 in-state students and 250 out-of-state students.

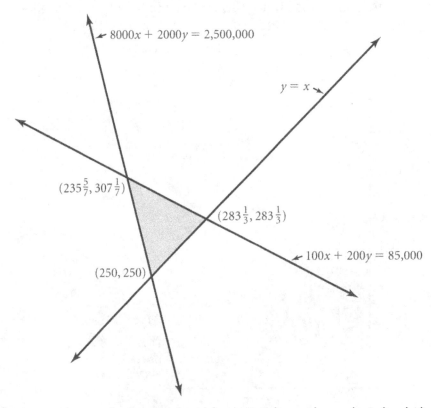

In their presentations, students should review how they sketched the graph, including how they determined which point minimized the cost and how they found its coordinates.

In explaining how they identified the point, students should apply the "family of parallel lines" concept in which the combinations of in-state and out-of-state students that give any particular cost form a straight line that "slides" when the cost is changed. They need to choose the "minimum position" of this family of parallel lines, or the line in this family that passes through the feasible region and produces the lowest cost. If they sketch one of these lines (for example, $7200x + 6000y = 1{,}440{,}000$), they will see the general direction of the lines in the family.

The cost increases as the line moves up and to the right, so the minimum cost is at the point $(250, 250)$.

Inventing Problems

Intent

This activity concludes *Cookies and the University* by asking students to create their own two-equation/two-unknown problems.

Mathematics

Real-world contexts have driven the mathematical activities throughout the IMP curriculum. In addition to learning how to solve equations and systems of equations, students have been learning how to set up equations by expressing conditions in real-world contexts symbolically. In this activity and the rest of the unit, students create their own contexts for problems involving systems of equations and then set up and solve them.

Progression

Students create problems individually and then work in groups to solve one another's problems. Each group presents its "best" problem and solution to the class.

Approximate Time

20 minutes for activity (at home or in class)

25 minutes for presentations

Classroom Organization

Individuals, then small groups, followed by class presentations

Doing the Activity

When introducing the activity, you might mention that "professional problem writers" who write textbooks or IMP units, for example, often come up with the general idea for a problem, write possible equations, and then go back and adjust the numbers in the problem so that the equations and solutions are reasonable.

Discussing and Debriefing the Activity

When everyone has created a problem, have students work in groups on one another's problems. Each group should then select its "best problem" to present to the class. Choose a student at random from each group to present the group's best problem and its solution.

Supplemental Activities

And Then There Were Three (extension) extends the ideas from *Get the Point* and *Inventing Problems* to systems of three linear equations in three unknowns.

An Age-Old Algebra Problem (extension) presents another situation for which students must set up and solve a system of three linear equations in three unknowns.

Creating Problems

Intent

Students complete their study of linear programming problems by creating and solving their own problems. In addition, they review their work over the entire unit to compile their unit portfolios.

Mathematics

A review of several of the linear programming problems from the unit will help students identify three common features: they all involve two variables, include a set of conditions that define a feasible region of solutions, and have a linear function to be maximized or minimized. Students create new contexts that lead to linear programming problems. This creative activity is designed to support students' understanding of the structure of such problems and the methods for solving them.

As students begin to assemble their portfolios, they are asked to review their work through two lenses: work that developed methods for solving systems of two-variable/two-equation linear systems and work that developed an understanding of linear programming in two variables.

Progression

Students will work in groups to create linear programming problems, solve the problems, and present them formally to the class. In addition, they will present their results on the final POW of the unit, compile their unit portfolios, and complete the unit assessments.

Ideas for Linear Programming Problems

Producing Programming Problems

Beginning Portfolio Selection

Just for Curiosity's Sake

Producing Programming Problems Write-up

Continued Portfolio Selection

Cookies Portfolio

Ideas for Linear Programming Problems

Intent

Students review several problem situations from the unit to clarify the main characteristics of linear programming problems.

Mathematics

The linear programming problems encountered by students in this unit share three features. They each have

- two variables
- a set of conditions, represented as a system of linear inequalities, that constrains the solutions
- a function to be maximized or minimized

By reviewing these common features, students will prepare for the final activity of the unit: creating their own linear programming problems.

Progression

Working individually, students review three of the problem contexts from the unit. They then work in groups to generate ideas for their own linear programming problems.

Approximate Time

10 minutes for introduction

25 minutes for activity (at home or in class)

15 minutes for group exploration

Classroom Organization

Individuals, then groups

Doing the Activity

Introduce this activity by reviewing the two types of problems students have been solving.

- Two-equation/two-unknown problems: In these problems, a pair of linear equations describes the situation. The task is to find the common solution.
- Maximizing (or minimizing) problems: These problems begin with several linear inequalities that define a feasible region. The task is to maximize (or minimize) some linear expression within that region. Students have used a geometric argument, involving a family of parallel lines, to find which intersection point they are looking for, and only at that stage did they solve a pair of linear equations. Tell

students that this type of problem is called a **linear programming** problem.

In the previous activity, *Inventing Problems,* students created problems of the first type. In the final activity of the unit, *Producing Programming Problems,* each group will make up a linear programming problem and present its solution to the class.

Today's activity will generate ideas for those problems. It explains in more detail what a linear programming problem is and asks students to examine some problems of this type to get a clearer sense of what they have in common and how they work.

After students have answered Questions 1, 2, and 3 on their own, have them convene in their groups to share their answers and then work on Questions 4 and 5.

Discussing and Debriefing the Activity

You may want groups to present their ideas for Questions 4 and 5 to the class. You might list their ideas on chart paper and post them so that students can refer to them when they begin work on *Producing Programming Problems.*

With each example students offer, focus on whether the thing being maximized or minimized depends linearly on the chosen variables. If not, try to elicit a way to change the example so that it does do so. Also, make sure the constraints are linear conditions on the selected variables.

Producing Programming Problems

Intent

In this culminating activity, students apply all they have learned about linear programming to invent situations of their own.

Mathematics

As in *Inventing Problems,* students are asked to step back and create their own contexts for problems—in this case, linear programming problems—and then set up and solve them. They must include all the steps they have learned that are essential to understanding, translating, and solving their problems, and they must present them to their classmates in a clear and coherent way.

Progression

Working in groups, students use their work on *Ideas for Programming Problems* to write and solve linear programming problems. They then prepare and give formal presentations to their classmates. They will also develop criteria and use them to assess the presentations.

Approximate Time

70 minutes for activity

90 minutes for presentations

Classroom Organization

Groups and individuals, followed by whole-class presentations

Doing the Activity

Tell students that this activity will be part of their unit portfolios, so it is essential that everyone do his or her own write-up, to be completed in *"Producing Programming Problems" Write-up.*

Set up a process for reviewing students' ideas before they get very far along. Examine their problem contexts in terms of both mathematical and subject-matter appropriateness.

You may want to have groups grade the presentations on this activity, including their own. If you choose to have peer grading, discuss how you will use these grades. Also discuss grading criteria, perhaps letting students develop their own criteria, such as those listed here.

- Creativity
- Mathematical elegance
- Clarity of presentation
- Participation of all group members

Post the grading criteria for students to refer to while they work on the problems and during the presentations.

Before presentations begin, you might want to have a short discussion of what makes an excellent presentation. You may want to film the presentations to give the activity added importance.

Discussing and Debriefing the Activity

Allow students a reasonable amount of time to evaluate the quality of each presentation and to write their evaluations before moving on to the next presentation.

Beginning Portfolio Selection

Intent

As with all unit portfolios, this task supports students' process of reflecting on key ideas from the unit and selecting appropriate artifacts for their portfolios.

Mathematics

This beginning selection asks students to focus on linear programming problems in the unit, describe the steps to solve such a problem, and identify three activities that helped them understand the process.

Progression

Students work individually on this activity.

Approximate Time

45 minutes for activity (at home)

5 minutes for discussion

Classroom Organization

Individuals

Discussing and Debriefing the Activity

Have a couple of volunteers read their descriptions of the process for solving linear programming problems.

Just for Curiosity's Sake

Intent

This activity gives students one final opportunity to reinforce their understanding of methods for solving systems of linear equations.

Mathematics

In Part II, students once again encounter an **inconsistent system** of equations, or one with no solutions. In the discussion, students explore symbolic cues for identifying inconsistent systems. An optional extension offers an example of a **dependent system**, in which any solution to one equation is also a solution to the other.

Progression

Students work individually on the activity as they complete their work on *Producing Programming Problems* and as they begin to compile their unit portfolios. They share ideas in groups and in a class discussion.

Approximate Time

20 minutes for activity (at home or in class)

15 minutes for discussion

Classroom Organization

Individuals, then groups, followed by whole-class discussion

Doing the Activity

This activity requires little or no introduction.

Discussing and Debriefing the Activity

Ask students to work in groups to check their answers for the systems in Part I.

In Part II, students should recognize that no pricing policy would give the stated results. Nevertheless, they should be able to write a pair of equations that describe the situation. Using R for the cost of a reserved-seat ticket and G for the cost of a general-admission ticket, the equations are

$$200R + 800G = 20,000 \quad \text{(for the first night)}$$

$$250R + 1000G = 23,000 \quad \text{(for the second night)}$$

Get students to articulate that these two equations have graphs that are parallel lines, which means they have no common solution.

Ask students how they might show symbolically that these two equations are inconsistent, perhaps using equivalent equations. How could you use equivalent equations to verify that these equations are clearly inconsistent? One approach is to divide the first equation by 200 and the second by 250, giving $R + 4G = 100$ and $R + 4G = 92$, respectively. Another is to note that the number of tickets sold of each type rose exactly 25 percent on the second night but that the money taken in did not.

Optional Extension of Part II

If time allows, pose this follow-up question.

When the inconsistency in the receipts was pointed out, the bookkeeper checked the records. Sure enough, the amount taken in on the second night was actually $25,000, not $23,000. What does that tell you about the prices of the two types of tickets?

Let groups explore this question for a few minutes. They should realize that they still can't determine the prices, because the new second equation,

$$250R + 1000G = 25{,}000$$

has the same graph as the first equation, $200R + 800G = 20{,}000$. (Both equations simplify to $R + 4G = 100$.)

Students should realize that any point on that graph will satisfy the equation (with perhaps the added condition that $R > G$, because reserved-seat tickets should cost more than general-admission tickets). But they should also realize that there is not enough information to decide exactly which point represents the actual prices. You can use this occasion to review the term **dependent equations**.

Key Question

How could you use equivalent equations to verify that these equations are clearly inconsistent?

Producing Programming Problems Write-up

Intent

In this activity, students prepare write-ups of their group's linear programming problems.

Mathematics

In their write-ups, students must include all the steps essential to understanding, translating, and solving their group's linear programming problems.

Progression

Students work individually to write a statement of the group's problem, the solution, and a proof that the solution is the best possible.

Approximate Time

30 minutes for activity (at home or in class)

Classroom Organization

Individuals

Doing the Activity

Remind students that their write-ups of the problems they created and solved will become part of their portfolios.

Continued Portfolio Selection

Intent

Students reflect on key ideas from the unit by selecting two problem situations that could be solved using a system of linear equations, to include in their portfolios.

Mathematics

In this activity, students focus on the theme of solving systems of linear equations with two variables.

Progression

Students work on this activity individually in anticipation of completing their unit portfolios.

Approximate Time

30 minutes (at home or in class)

Classroom Organization

Individuals

Cookies Portfolio

Intent

Students compile their unit portfolios and write their cover letters.

Mathematics

Students' portfolio selections will include the artifacts they identified in *Beginning Portfolio Selection* and *Continued Portfolio Selection,* which focus on the two types of problems explored in the unit: systems of linear equations in two variables and two-variable linear programming problems.

Progression

Students start work on their portfolios in class by reading the instructions in the student book. They then work independently to review their work in the unit, select samples, reflect on the evidence of their learning, and write cover letters.

Approximate Time

10 minutes for introduction

40 minutes for activity (at home)

Classroom Organization

Individuals

Doing the Activity

Have students read the instructions in the student book carefully.

Discussing and Debriefing the Activity

You may want to have students share their portfolios in their groups, comparing what they wrote about in their cover letters and the activities they selected.

Profitable Pictures

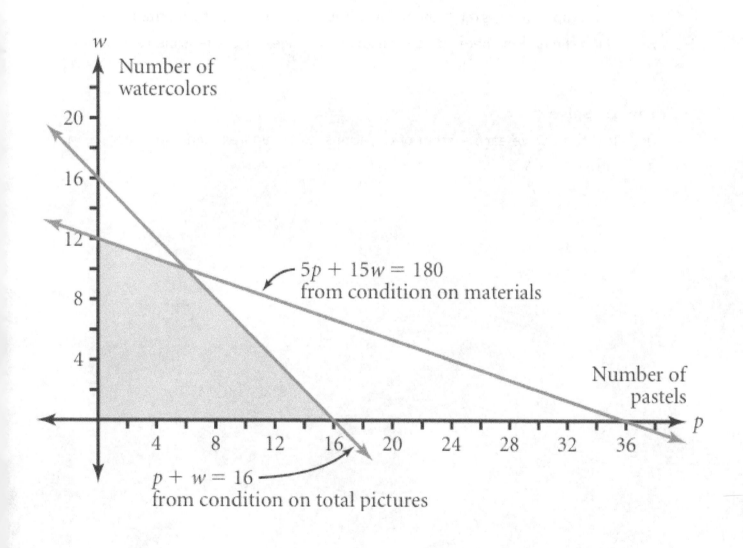

In-Class Assessment

Part I: Graph It

Consider the following constraints.

$$x \geq y$$
$$x + y \geq 50$$
$$6x + 5y \leq 500$$
$$x \geq 0, y \geq 0$$

1. On graph paper, sketch the feasible region for this set of constraints.

2. Find the approximate coordinates of each vertex of the feasible region.

Part II: Solve It

Use algebra to solve each system of equations. Show and explain your work clearly.

3. $4x + 3y = 5$

    $2x - 5y = 9$

4. $4x - 6y = 20$

    $6x - 9y = 24$

Take-Home Assessment

Part I: What If . . . ?

These three problems are variations on the unit problem. Your task is to find the combination of plain and iced cookies that maximizes the Woos' profit in each new situation. Consider these three variations as three separate problems.

Each problem has a graph that shows the feasible region of the original problem. The shaded area represents that original feasible region.

Questions 1 and 2 show a profit line based on the original problem. Question 3 shows a profit line based on a different profit expression.

The other lines in each problem are the graphs of the original problem's constraint inequalities. Here are those inequalities.

$$P + 0.7I \leq 110 \qquad \text{(for the amount of cookie dough)}$$
$$0.4I \leq 32 \qquad \text{(for the amount of icing)}$$
$$P + I \leq 140 \qquad \text{(for the amount of oven space)}$$
$$0.1P + 0.15I \leq 15 \qquad \text{(for the amount of the Woos' preparation time)}$$

1. Suppose everything is the same as in the original problem, *except* that the Woos have an unlimited amount of dough. What combination of plain and iced cookies will maximize profit? Explain your answer.

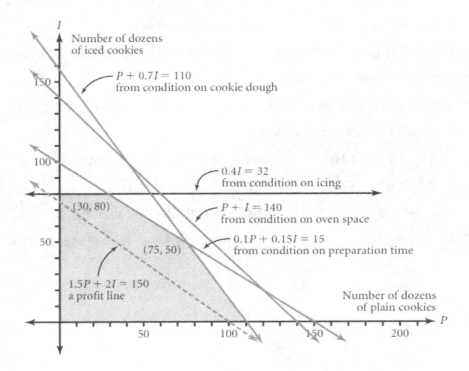

2.   Suppose everything is the same as in the original problem, *except* that the Woos have an additional constraint. They can't sell more than 60 dozen plain cookies. What combination of plain and iced cookies will maximize profit? Explain your answer.

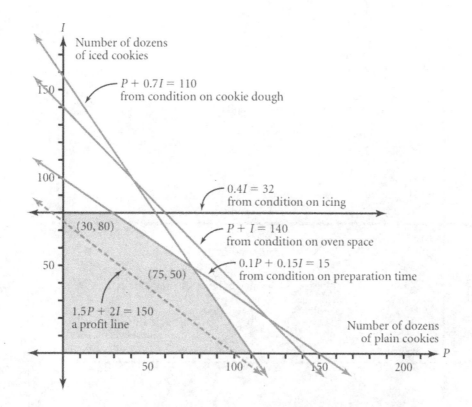

3.   Suppose everything is the same as in the original problem, *except* the profit on each kind of cookie. The Woos make a profit of $2.00 on each dozen plain cookies and $4.00 on each dozen iced cookies. (The original profits were $1.50 and $2.00.) What combination of plain and iced cookies will maximize profit? Explain your answer.

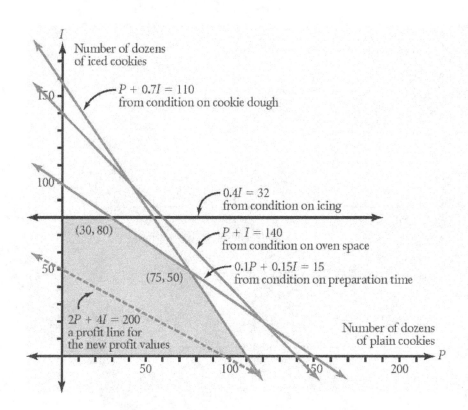

Part II: The Pebbles Rock at Big State U

The Rocking Pebbles are playing a concert at Big State University. The auditorium seats 2200 people. The concert manager decides to sell some tickets at $10 each and the rest at $15 each. How many of each kind should the manager sell if he wants ticket sales to total $26,600? Assume that all the tickets will be sold.

Find the answer by setting up and solving a system of two linear equations with two unknowns. Show and explain your work clearly.

Cookies Guide for the TI-83/84 Family of Calculators

This *Calculator Guide* gives suggestions for calculator use with selected activities of the Year 2 unit *Cookies.* The associated *Calculator Notes* that you download contain specific calculator instructions that you might distribute to your students. NOTE: If your students have the TI-Nspire handheld, they can attach the TI-84 Plus Keypad (from Texas Instruments) and use the calculator notes for the TI-83/84.

In the first part of the *Cookies* unit, students will learn to graph systems of linear inequalities by hand and may use the calculator very little. As they begin to maximize a function within the feasible region, the issue of solving a pair of linear equations for the point of intersection becomes important. After the activity *A Rock 'n' Rap Variation,* students are introduced to how to do this on the graphing calculator. Algebraic solutions often provide greater accuracy, but students will have many opportunities to check their algebra by graphing. Although checking by graphing would be tedious on paper, it is very practical on the graphing calculator.

Students who enjoy exploring the calculator's capabilities will find the **Shade(** command to be an interesting way to shade a feasible region on the calculator. They can learn how to use this by referring to the calculator's manual or the Internet.

Profitable Pictures: As students are sharing their methods for finding the point of intersection of the two lines in this activity, they may suggest that a more accurate estimation could be made by using a graph of the equations made on a graphing calculator. If students bring this up, ask what they would need to do to the equations to enter them as functions on the Y= screen. (They'd need to solve for one of the variables.) If students don't bring this up at this time, don't push it. The activity *Getting on Good Terms* reviews the necessary algebra for solving the equations for a particular variable. Graphing the feasible region on the calculator is introduced following the activity *A Rock 'n' Rap Variation.*

A Rock 'n' Rap Variation: After students have finished the activities *Rock 'n' Rap* and *A Rock 'n' Rap Variation,* lead a class exploration of using a graphing calculator to solve this problem. The details of how to lead the discussion are described in the *Teacher's Guide* notes for this activity. See the *Calculator Note* "Solving a Linear Programming Problem" for keystroke instructions on how to do this; you might distribute this *Calculator Note* to students.

Getting on Good Terms: This activity reviews the process of solving for *y,* which is needed for an equation to be entered into the Y= screen on a calculator, but a calculator is not required for the activity. However, notice the issue of decimal approximations discussed in the *Teacher's Guide*—students may solve the equation $5x + 3y = 17$ and use decimal approximations of fractions to get $y = 5.67 - 1.67x$ (rather than $y = (17/3)$—

(5/3)*x*). This may lead to slightly inaccurate graphs or approximate solutions. Students who are uncomfortable with fractions may need to be reminded of this issue several times.

Get the Point: Students may be tempted to use graphing to solve the equations in *Get the Point.* Clarify that the purpose of this activity is to develop the ability to solve the equations algebraically, which will yield exact answers instead of approximations. Encourage students, however, to use graphing as one way to quickly check that their answers are reasonable. They should also become comfortable with checking by substitution, so you may want to assign different methods for checking particular problems.

Only One Variable: Although you will certainly want students to solve Question 1 of this activity algebraically, point out again how the problems might be checked by graphing each side of the equation as a separate function and looking for the intersection. Students often overlook this method of solving single-variable equations.

Set It Up: When discussing this activity, you may want to assign one group to use the overhead projector calculator to present the solution graphically, in addition to the groups that are presenting algebraic solutions and sketched graphs.

A Reflection on Money: Again, when discussing this activity, you may want to assign one group to use the overhead projector calculator to present the solution graphically, in addition to the groups that are presenting algebraic solutions and sketched graphs.

More Linear Systems: It is very useful to have students graph the inconsistent system of equations in Question 2c, and then also graph a dependent system in the follow-up discussion. In this early stage of students' algebraic development, they may decide too quickly that "there is no solution" when they are having trouble arriving at a solution that will check. Remind them that graphing gives them a quick way to check whether their suspicions are true.

In-Class Assessment: The in-class assessment for this unit will definitely require you to thoroughly clear the calculators between class periods. Clearing the home screen using the CLEAR key does not erase the functions entered on the Y= screen—you'll need to also press Y= and delete each function by moving onto it and pressing CLEAR. The calculator can be cleared more quickly by resetting the memory, but note that this will also clear all programs and set the calculator to radian mode. If that's not a problem, reset the memory using these keystrokes on the TI-83/84: 2ND [MEM] 7 1 2.

Calculator Notes

Solving a Linear Programming Problem

This note describes how to use a calculator to solve a linear programming problem, using as an example the problem from the activity *Rock 'n' Rap.*

Begin by pressing Y= and entering the equations of the three boundary lines of the feasible region. You'll first have to rewrite the equations $15,000x + 12,000y = 150,000$ and $18x + 25y = 175$ in "$y =$" form. (If you didn't use x and y in your original equations, replace them now. Note that either variable could be x or y, as neither equation is clearly independent or dependent.) Be sure to use parentheses as necessary to control the order of operations.

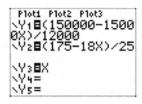

Press GRAPH to view the graph. Adjust the viewing window as needed. To do this, press WINDOW, and adjust the values of **Xmin, Xmax, Xscl, Ymin, Ymax,** and **Yscl.** Think about what the variables represent, and therefore what numbers are reasonable to use for the minimum and maximum. Are you interested in negative numbers of CDs?

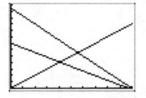

[0, 10, 1, 0, 13, 1]

Think about which area of the graph represents the feasible region. The overlapping triangular region in the graph at right shows the feasible region. This graph was created using the **Shade(** command. To learn more about how to use this command, consult your calculator manual or the Internet.

Next, graph parallel lines representing different profits. Again, you will have to rewrite the equations in "$y =$" form. For example, the profit equation $20,000x + 30,000y = 120,000$ becomes $y = (120,000 - 20,000x)/30,000$ or the equivalent. In this illustration, the profit line is graphed in a heavy line style. To do this, move your cursor to the left of Y4 in the Y= screen, and press ENTER until you see the heavy line icon.

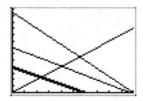

Now adjust the profit line by substituting a different number for 120,000 in the equation. Find what maximum profit is possible while keeping some part of the profit line within the feasible region.

Try using the Trace feature to find the coordinates of the desired point. Press TRACE, use the up and down arrows to move onto a different line, and use the left and right arrows to move the cursor back and forth on a line.

CPSIA information can be obtained
at www.ICGtesting.com
Printed in the USA
BVHW011031300421
606029BV00007B/80